INTEGRABLE SYSTEMS IN STATISTICAL MECHANICS

SERIES ON ADVANCES IN STATISTICAL MECHANICS*

ISSN: 2010-1996

Series Editor: Davide Cassi *(University of Parma, Italy)*

Published

Vol. 20: *Non-equilibrium Thermodynamics of Heterogeneous Systems Second Edition*
by S. Kjelstrup & D. Bedeaux

Vol. 19: *An Introduction to Stochastic Processes and Nonequilibrium Statistical Physics*
by H. S. Wio, R. R. Deza & J. M. López

Vol. 18 *Statistical Mechanics of Magnetic Excitations: From Spin Waves to Stripes and Checkerboards*
by E. Rastelli

Vol. 17: *Chaos: From Simple Models to Complex Systems*
by M. Cencini, F. Cecconi & A. Vulpiani

Vol. 16: *Non-Equilibrium Thermodynamics of Heterogeneous Systems*
by S. Kjelstrup & D. Bedeaux

Vol. 15: *Lattice Statistics and Mathematical Physics: Festschrift Dedicated to Professor Fa-Yueh Wu on the Occasion of his 70th Birthday*
edited by J. H. H. Perk & M.-L. Ge

Vol. 14: *Statistical Physics on the Eve of the 21st Century: In Honour of J. B. McGuire on the Occasion of his 65th Birthday*
edited by M. T. Batchelor & L. T. Wille

Vol. 13: *Exactly Soluble Models in Statistical Mechanics: Historical Perspectives and Current Status*
edited by C. King & F. Y. Wu

Vol. 12: *Quantum Many-Body Systems in One Dimension*
by Zachary N. C. Ha

Vol. 10: *An Introduction to Stochastic Processes and Nonequilibrium Statistical Physics*
by H. S. Wio

Vol. 9: *Disorder and Competition in Soluble Lattice Models*
by W. F. Wreszinski & S. R. A. Salinas

*For the complete list of titles in this series, please go to
http://www.worldscientific.com/series/sasm

Series on Advances in Statistical Mechanics - Volume 1

INTEGRABLE SYSTEMS IN STATISTICAL MECHANICS

Editors

G M D'Ariano

A Montorsi

M G Rasetti

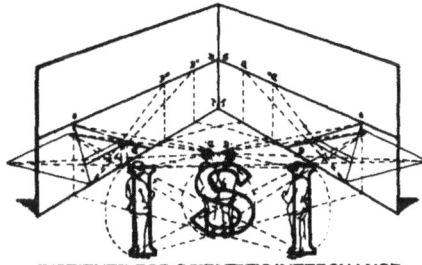

INSTITUTE FOR SCIENTIFIC INTERCHANGE

World Scientific

Published by

World Scientific Publishing Co. Pte. Ltd.
5 Toh Tuck Link, Singapore 596224
USA office: 27 Warren Street, Suite 401-402, Hackensack, NJ 07601
UK office: 57 Shelton Street, Covent Garden, London WC2H 9HE

British Library Cataloguing-in-Publication Data
A catalogue record for this book is available from the British Library.

Series on Advances in Statistical Mechanics — Vol. 1
INTEGRABLE SYSTEMS IN STATISTICAL MECHANICS

ISBN-13 978-9971-978-11-2
ISBN-10 9971-978-11-3
ISBN-13 978-9971-978-14-3 (pbk)
ISBN-10 9971-978-14-8 (pbk)

FOREWORD

This is the first book produced as output of one of the activities promoted by the Institute for Scientific Interchange I.S.I. in Torino, Italy. The I.S.I. is an institution, supported primarily by the Regional Government of Piedmont, devoted to the encouragement, support and patronage of science, in the broadest, indifferentiated sense of the word. It is meant to provide the scientific community both a direct intellectual support and a source and reservoir of high quality forces through the promotion of interchange activities at different levels, bridging those gaps which are unavoidable in the traditional modalities of the international scientific cooperation.

Its primary purpose is to bring together a small number of scholars, working on a variety of related problems, so as to provide the visiting members with an exacting ambient where they could feel however their own intellectual development and growth as one of the principal purposes, and the local community with an accessible environment as lively and stimulating as rich and fertile, comparable with the most advanced similar institutions in the world.

The inevitable (and desiderable) limitation on the size of the institution programs naturally promotes the tendency to focus, with special emphasis, on disciplines and topics which have in the Region and the City some confirmed tradition of fruitful activity. The scientific programs are designed to have as one common characteristic continuity in time, attendance and disciplines, and are based on the presence in Torino of distinguished scientists, visiting for variable periods of time, to give sets of lectures, conduct seminars and lead research projects.

Together with them a group of scientists active in the same field, desirably young promising researchers, take part in the seminars, both providing an active audience and acting as the medium of a proficuous interaction between the scholars and the whole scientific community.

M. Rasetti
T. Regge

CONTENTS

INTRODUCTION

The number of exactly solved models in the Statistical Mechanics of classical two-dimensional systems has grown to be relatively large. The field has had its successes uniformly distributed in the period between 1944, when Lars Onsager so ingeniously solved the two-dimensional Ising model, and 1980 when Rodney Baxter produced the solution of the hard hexagon model. In between ice-type models, vertex models and some related models, such as the Ashkin-Teller, Potts or the hard square models, were either solved, or partially dealt with, very often by not so different techniques. Thus recently, the attention of mathematical physicists has been focused more and more on the connection between the different methods of solution, in the effort to understand the relevant underlying structure and to try to extend it eventually to higher dimensional systems. Also the equivalence of several of such classical models with quantum ones, living in one-dimensional space, has played a relevant role, raising the hope to find in the methods characteristic of quantum mechanics and field theory a way out of the often formidable difficulties.

Finally the exact solvability has been related, or – better – people have tried very hard to connect it, with integrability in the customary sense of mechanics; namely with the existence, in the thermodynamic limit, of infinitely many conserved quantities. In the latter context, new dinamical features such as the existence of soliton-type solutions have entered the picture. Conventionally, one talks about exactly solved model whenever some physically significant quantity, such as the free energy or an order parameter or some correlation function have been worked out analytically into a mathematically closed manageable form, or at least their calculation has been reduced to a problem of classical analysis – such as the evaluation of an integral or the solution of a differential (or integral) equation etc – . There is plenty of discussions going on about the physical relevance of such solutions, indeed of the models themselves. Some claim that they are unphysical to the point of being useless. Others believe that the mathematical tools one has to resort to in order to find these solutions are far too complicated to justify the effort. The fact is that on the one hand some of these problems have induced as a feedback a wealth of good results in mathematics (such as the theory of infinite dimensional Lie algebras, the combinatorial topology, the theory of non-linear differential equations). On the other hand even when not strictly corresponding to realistic systems, the exact solution of the models has permitted to check more general theorems and conjectures in statistical physics (such as the Lee-Yang theorem in the theory of phase transitions for which the solution of the two-dimensional Ising model has played a crucial role of reference; or the concept of universality of the critical exponents, which has been usefully forced by the results of Lieb and Baxter; or, yet, the famous Bethe Ansatz, which has allowed a deeper comprehension of quantum excitations in one dimensional systems and has lead recently to the exact solution of the Kondo problem).

Finally both classical two dimensional and quantum one dimensional systems do exist in the real world; and the agreement between the output of the models and those of the measurements in the laboratories is very often exceptional.

The reduction of the different methods of solution to a unique global general method has also had unespected successes. The most instructive instance is that of the ice model, of different ferroelectric models, of the discrete lattice gas or the Ising model; all included as particular cases in the eight vertex model, defined by Fan and Wu and solved, in its whole generality, by Baxter.

The logical steps lead from the dimer covering technique, through the introduction of Grassmann variables, to the Pfaffian method, and from the latter, by the very general procedure of the transfer matrix, to a quantum one dimensional problem.

Thus the equilibrium configurations of classical two-dimensional systems with nearest neighbour interactions were shown to exhibit a strict topological analogy with the space-time trajectories of a dynamical system in one spatial dimension. It is a somewhat simple, though surprising, extension of an idea due to C. N. Yang, exploited when he was dealing with the thermodynamics of a one dimensional chain of hard-core bosons, the unbelievably deep fact that the integrability of the transfer matrix (and that of the dynamical system) is connected with the existence of ternary relations among micro-transfer-matrices. Such relations are remarkable realizations of subgroups of the permutation group (braids), and imply the existence of one-parameter families of mutually commuting transfer matrices (whereby the integrability follows). Also Yang's method foreshadows the inverse scattering technique thoroughly developed by Faddeev, which links the symmetry conditions with those of periodicity, beautifully setting in operatorial form Bethe's Ansatz. It is therefore in a way no more surprising that Zamolodchikov's factorization equations for the S-matrix, representing the conditions which are necessary to factorize a multi-particle scattering matrix into two-particle ones, coincide with Yang's equations and Baxter's "triangle" equations.

This book stems out of a set of lectures delivered in Torino (with the exception of Zamolodchikov, who could not attend, but submitted one of his papers, to be reprinted as his contribution to the collection) in the spring and summer of 1984. Organized by I.S.I. the lectures were meant to review the state of art on the subject of integrable systems in Statistical Mechanics, both classical and quantum, and to give a thorough updating on the most recent results as well as perspectives in the field. The lectures were taped, then rewritten by the editors and finally submitted to the authors, to whom the editors want to express their gratitude for the careful and patient correction of the manuscript. The accurate and beautiful typing is due to the passionate and dedicated work of Andrea Leone to whom the editors convey their deepest thanks. Finally, we acknowledge Csi–Piemonte for making available the TEX system with which the typing was done.

Torino, january 1985

Giacomo D'Ariano
Arianna Montorsi
Mario Rasetti

Exactly Solved Models in Statistical Mechanics

RODNEY J. BAXTER F.R.S.

Department of Theoretical Physics
Research School of Physical Sciences
The Australian National University
Canberra, A.C.T., Australia

INTRODUCTION

The first exact solution in statistical mechanics of one-dimensional models was for a hard core gas ("Tonks" gas)[1]; Ising then solved in 1925 the famous "Ising" model [2]. Later on a "hard core with exponential attraction" model (HCEA) was solved by Takahashi [3].

Unfortunately all these models have no phase transitions, according to Van Hoove [4] theorem (1950) which says that no phase transitions can occour in short range one-dimensional models at a non-zero temperature, unless the interaction involves infinitely many particles.

Hovever the "infinite dimensional" models, as the mean-field one [5] and the Bethe one [6] can be exactly solved showing phase transitions.

Here it's useful to define the dimensionality d of a lattice as the exponential of the power law N^d describing the asymptotic growth of the number of N-neighbours of a certain site visited in N steps; the system is infinite dimensional when the number of neighbours visited grows faster than any d.

In the mean-field model every particle interacts equally with all the others.

The Bethe model is defined on a lattice which is the Cayley tree infinitely far from the boundary, in the termodynamic limit $N \to \infty$. A Cayley tree is the graph shown in fig. 1.

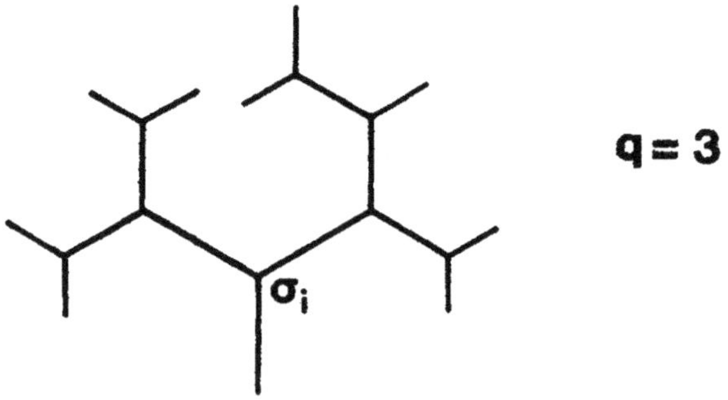

fig. 1

One starts with a certain site (i) and connects it with q sites; then each of these q sites is connected with $(q-1)$ sites, and so on. Such a graph contains no circuits. The same holds for the Bethe lattice which is deep within the Cayley graph and can be thus thought of as a lattice of coordination number q.

The HCEA model as well is an infinite dimensional model exhibiting phase transitions, if one solves it in the limit of weak and infinitely long-range attraction [7].

All these infinite dimensional models satisfy the scaling hypotesis and have classical exponents (see following).

This is not completely true for the spherical model (solved by Berlin and Kac in 1952 [8]), which is not infinite dimensional but involves essentially infinite range interactions and has phase transitions for $d > 2$. In three dimensions however the exponents are not classical, while they are for $d \geq 4$. The two dimensional lattice models that have been solved are few; in these lectures we'll discuss mainly these.

Generally they have phase transitions with exponents which are not classical. These models are also interesting because they describe real systems, as thin films or crystals with anisotropic interactions.

It's worth specifying that by the phrase ***exactly solved*** we don't mean ***rigorously solved.*** Infact ***exactly*** applies to the results, while ***rigorously*** applies to the way you get it. Exact results can be obtained by an applied mathematician derivation which is not properly a rigorous proof. For instance one can multiply and diagonalize infinite matrices without demonstrating that matrix products are convergent, believing that this is true and that the results are exactly correct.

Exact results can also be obtained by a conjecture; for example De Gaunt in 1967 obtained by a guess the correct value of the critical activity of a hard exagon model and this value was then verified in 1979 by Baxter [9].

In table 1 are listed the main exactly solved models in two dimensions:

MODEL	SOLVED
Ising	Onsager, 1944
Dimers	Kasteleyn and Fisher, 1961
6-Vertex KDP	Lieb, 1967
Vertex F	Lieb, 1967
8-Vertex	Baxter, 1971
3 Spin	Baxter and Wu, 1980
Hard Hexagons	Baxter, 1980

table 1

The Ising model in two dimensions was solved by Onsager in 1944 [10].

Kasteley and Fisher solved indipendently the dimers problem in 1961 [11, 12]; it then turned out to be equivalent to the Ising model. In the dimers problem there is a lattice on which one can put dimers, i.e. objects that cover two sites at a time (like diatomic molecules), with the constraint that they are hard-core and so one site cannot be covered by more than one dimer; the problem is to count how many ways there are of covering the lattice with such dimers.

The dimer covering is an alternative approach to the Onsager solution of the Ising model; the fundamental trick consists in the one to one correspondence between the dimer covering and the elements of the Pfaffian of an antisymmetric matrix.

The six-vertex model was solved in 1967 by Lieb [13, 14, 15]. He solved three specific cases: ice type, KDP and F models. The ice model is a special case of the F model.

Six-vertex model are all specific cases of the more general eight-vertex model, solved by Baxter in 1971 [16, 17]; the 8-vertex model contains as a special case also the 3-spin model [18].

Finally a recent result is that even the hard exagon model can be identified as an eight-vertex model; it can infact be recognized as a six-vertex decorated model.

1. CRITICAL EXPONENTS

For all these models we calculate of course the free energy, but we are also interested in their critical exponents (table 2) to describe their behaviour around the critical point:

MODEL	$\bar{\mu}$	α	β	μ	δ
Ising	$\frac{\pi}{2}$	0	$\frac{1}{8}$	1	15
Dimers	$\frac{\pi}{2}$	0	$\frac{1}{8}$	1	15
6-Vertex KDP	π	1			15
6-Vertex F	0	$-\infty$	∞	∞	15
8-Vertex	$0-\pi$	$2-\frac{\pi}{\bar{\mu}}$	$\frac{\pi}{16\bar{\mu}}$	$\frac{\pi}{2\bar{\mu}}$	15
3-Spin	$\frac{3\pi}{4}$	$\frac{2}{3}$	$\frac{1}{12}$	$\frac{2}{3}$	15
Hard Hexagons	$\frac{3\pi}{5}$	$\frac{1}{3}$	$\frac{1}{9}$	$\frac{5}{6}$	14

table 2

Typically in these models there are temperature (T) and field (H) variables. The free energy is a function of them; generally one finds that it is an analitic function of H and T except for the line $H = 0$ and T from zero to T_c in the (T, H) plane. Really the first derivative of the free energy has a discontinuity crossing this cut, but the very significant singularity is at the critical point ($H = 0, T = T_c$).

At this purpose one may note that for example the Onsager's solution of the Ising model was just along the line $H = 0$, and that is not too much useful because this line is not typical of the whole plane at all. And infact it is not typical but it is the most interesting line one can look at.

Anyway the two dimensional Ising model with $H \neq 0$ is still unsolved: we know only numerically it's behaviour off the axis and in part this is due to the fact that we know the exact solution on the line $H = 0$.

So, if one fixes $H = 0$ and looks at the free energy f as a function of T, f normally shows a singularity in $T = T_c$ which goes like a power $(2 - \alpha)$ of $(T - T_c)$; note that if α is positive then it characterizes the divergence exponent of the specific heat (i.e. the second derivative of the free energy).
Similarly, looking at the spontaneous magnetization, one gets another exponent β of a function vanishing coming from below to the critical point.

One can also define the interfacial tension, which is the contribution to the free energy from two ordered domains in contact with one another.

Actually the last exponent δ is defined just along the line $T = T_c$; as the magnetization is a function of H, so H is a function of the magnetization.

One can test on table 1 the scaling hypotesis, which is really built in the renormalization group theory; in two dimensions the scaling predictions are:

$$2\mu = 2 - \alpha$$

$$\delta = -1 + (2 - \alpha)/\beta$$

In particular for the hard exagon model μ is obtained using these predictions, so as δ for almost all models.

Before going further let's say something about what such exponents are for these solved models (table 2).

Onsager found for the Ising model that α is zero: this just means that the specific heat diverges logaritmically. In this model δ is obtained independently from the scaling predictions, so that the scaling for δ has been actually tested on the Ising model.

For the KDP model $\alpha = 1$ corresponds to a first order transitions, while it is a bit harder to define β and μ because the system is frozen in it's ordered phase.

The F model has an infinitely weak singularity for the free energy: all its derivatives exist and are continuous but nevertheless the function is not analytic.

Before the solution of the eight-vertex model it was generally believed that all exponents of all models were universal.

Universality says that critical exponents shouldn't depend on the details of interaction. Of course they will depend on something, specifically on the dimensionality and on the symmetries of the system. That's certainly true for the Ising model: if we take this model with interaction in both horizontal and vertical directions and make those interactions different, we don't change the critical exponents at all. Besides we know that for the Ising model $H = 0$ is a special symmetry line, for which there is a critical point at $T = T_c$: out of the cut-line $H = 0$, $0 \leq T \leq T_c$ in the (H, T) plane, we have no singularity.

Universality is a very actractive idea because it means that real interactions can be modeled by very simple ones if we just want to know critical exponents: for example knowing the critical exponents of the three dimensional Ising model one would know them for an alloy such as bronze. Nevertheless when we look at the eight-vertex model, we have to introduce a parameter $\bar{\mu}$ (laying between 0 and π) which is just defined by the details of the interaction: varying $\bar{\mu}$ means to vary α, β and μ (not δ). It is now appreciated that there are indeed such models in which critical exponents are not universal: even the square lattice Ising model with ferromagnetic nearest and antiferromagnetic next-nearest neighbours interaction has non universal exponents.

The three spin model as well as the hard exagon model are special cases of the 8-vertex model corresponding respectively to $\bar{\mu} = \frac{3}{4}\pi$ and $\bar{\mu} = \frac{3}{5}\pi$. It is now possible to draw a sort of $\bar{\mu}$ line and consider the various model on it (fig. 2):

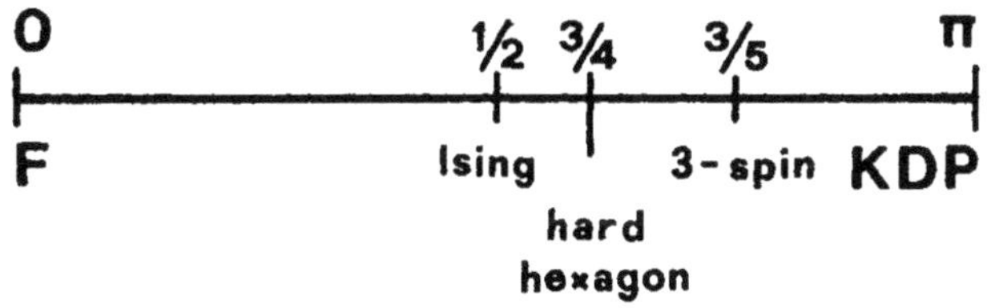

fig. 2

Interactions round a face model

All these exactly solved two dimensional models can be defined as special cases of a very general model, called the ***interaction round a face*** model (IRF).

Take a square lattice of N sites (fig. 3); on each site (i) put a spin σ_i which has some set of possible values (for most of these models $\sigma_i = +1$ or -1, while for hard exagons it's more natural to put $\sigma_i = +1$ or 0).

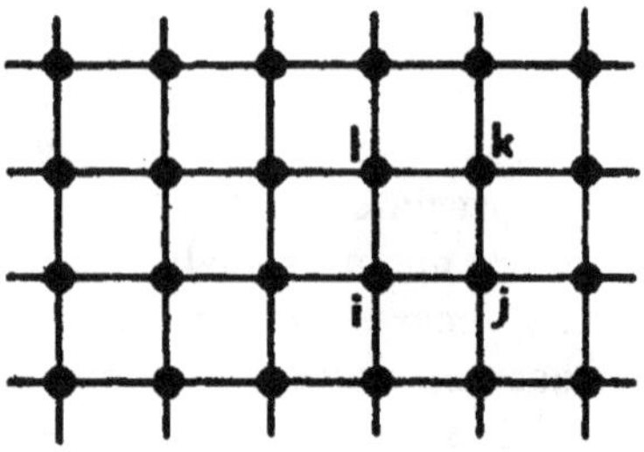

fig. 3

For the whole lattice one defines a hamiltonian which is a sum over all faces of the lattice of a face energy being some function ϵ of the four spins around that face:

$$H = \sum_{\substack{all \\ faces}} \epsilon(\sigma_i, \sigma_j, \sigma_k, \sigma_l) \tag{1.1}$$

Then the partition function is:

$$Z_N = \sum_{\sigma_l} \cdots \sum_{\sigma_N} e^{-\beta H} \qquad \beta = \frac{1}{K_B T} \tag{1.2}$$

If each spin has two possible values, then there are sixteen possible values for the argument of the function ϵ, so there are sixteen degrees of freedom; actually some of them are redundant because if we add an energy to one side of the square face and subtract the same energy from the other side we don't change anything at all.

One can define a Boltzmann weight function of a face $w(a, b, c, d)$, where a, b, c and d are the four spins around that face:

$$w(a, b, c, d) = \exp\left\{\frac{-1}{K_B T}\epsilon(a, b, c, d)\right\}$$

Then (1.2) becomes:

$$Z_N = \sum_{\sigma_l} \cdots \sum_{\sigma_N} \prod_{\substack{faces \\ (i,j,k,l)}} w(\sigma_i, \sigma_j, \sigma_k, \sigma_l) \tag{1.3}$$

For our next calculations we are also interested in the free energy per site f,

$$f = -K_B T \lim_{N \to \infty} N^{-1} \ln Z_N \tag{1.4}$$

and in averages, where for example the average of a single spin σ_i sitting somewhere in the middle of the lattice is:

$$< \sigma_1 >= Z_N^{-1} \sum_{\sigma_i} \cdots \sum_{\sigma_N} \sigma_1 \prod_{faces} w(\sigma_i, \sigma_j, \sigma_k, \sigma_l) \tag{1.5}$$

It is certainly true that all the models we have described before are special cases of this IRF model. Let us see it more precisely.

Nearest neighbour Ising model

In that case the hamiltonian is:

$$H = -J \sum_{\substack{horizontal \\ edges \\ (i,j)}} \sigma_i \sigma_j - J' \sum_{\substack{vertical \\ edges \\ (j,k)}} \sigma_j \sigma_k \tag{1.6}$$

where $\sigma_1, \ldots, \sigma_n = +1$ or -1.

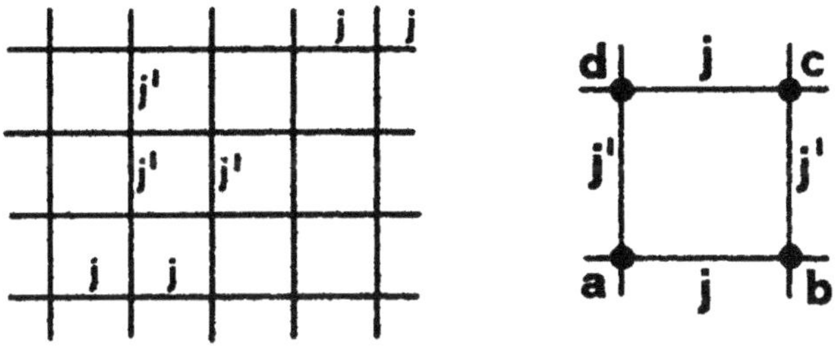

fig. 4

The minus sign before the interaction coefficients J, J' means we have ferromagnetic interaction. Infact with these signs, when two spins have the same value the energy is negative, when they have opposite signs it's positive; as negative energy is preferred to positive one, the system prefers equal spin.

That's the nearest-neighbour Ising model that Onsager solved; it can be incorporated into a more general IRF model by merely sharing out edge energies between adjacent faces:

$$\epsilon(a, b, c, d) = -\frac{1}{2} J(ab + cd) - \frac{1}{2} J'(bc + ad) \tag{1.7}$$

Note that this is not one of the special cases solved by eight-vertex model, while next nearest-neighbour Ising model is.

Next nearest neighbour Ising model

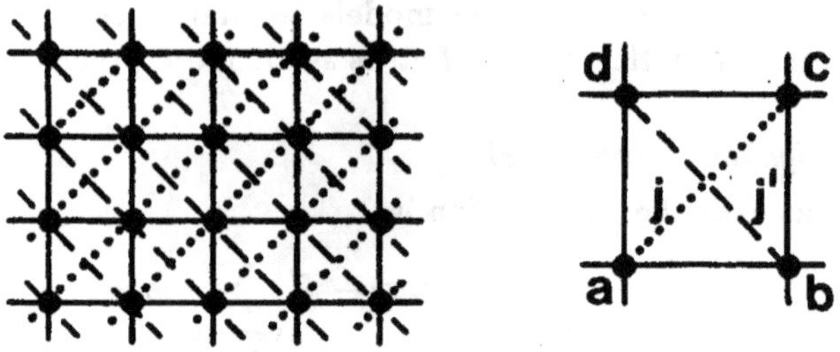

fig. 5

The hamiltonian is:

$$H = -J \sum_{\substack{NE-SW \\ diagonals \\ (i,k)}} \sigma_i \sigma_k - J' \sum_{\substack{NW-SE \\ diagonals \\ (j,l)}} \sigma_j \sigma_l \tag{1.8}$$

where again $\sigma_1, \ldots, \sigma_N = +1$ or -1.

This model can easily be recognized as an IRF model, the interaction being all inside the square:

$$\epsilon(a,b,c,d) = -Jac - J'bd \tag{1.9}$$

The next-nearest-neighbour Ising model factors into two independent and equal nearest-neighbour models. Infact *dot spins* interact diagonally only with nearest neighbours *dot spins* and the same is true for *cross spins*. So for the free energy for site we have:

$$f_{next\,n\,n}(J, J') = f_{n\,n}(J, J')$$

Eight-Vertex model

In next nearest neighbour Ising model the hamiltonian has two sublattice spin reversal symmetries. The most general model which has those symmetries and interactions only within a face is the eight-vertex model.

Obviously it's an IRF model:

$$\epsilon(a,b,c,d) = -Jac - J'bd - J_4abcd \tag{1.10}$$

The six-vertex model can be thought as a special limiting case of this model in which $-J, J', -J_4$ tend to infinite, the appropriate differences remaining finite. Under this limit, twelve of the possible values of $w(a,b,c,d)$ remain finite and non zero while the other four go to zero.

Three-spin model

It is defined over a triangular lattice which actually can be drawn as a square lattice with one set of diagonals (fig. 6):

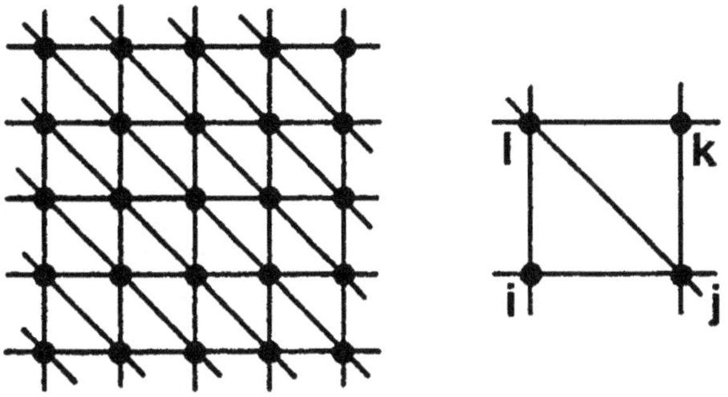

fig. 6

So there are two types of triangular faces (upper and lower). The hamiltonian is then:

$$H = -J \sum_{\substack{all\\ triangular\\ faces\\ (i,j,l)\, and (j,k,l)}} \sigma_i\sigma_j\sigma_l = -J \sum_{\substack{squares\\ i,j,k,l}} (\sigma_i\sigma_j\sigma_l + \sigma_j\sigma_k\sigma_l) \tag{1.11}$$

Again this is an IRF hamiltonian, with:

$$\epsilon(a,b,c,d) = -J(abd + bcd) \tag{1.12}$$

As formulated, the three-spin model is quite different from the ordinary Ising model. Infact there the hamiltonian is left unchanged by reversing all the spins while that's no more true for this model. Nevertheless we have a stronger symmetry, as can be seen dividing the triangular lattice into three equivalent sublattices (fig. 7):

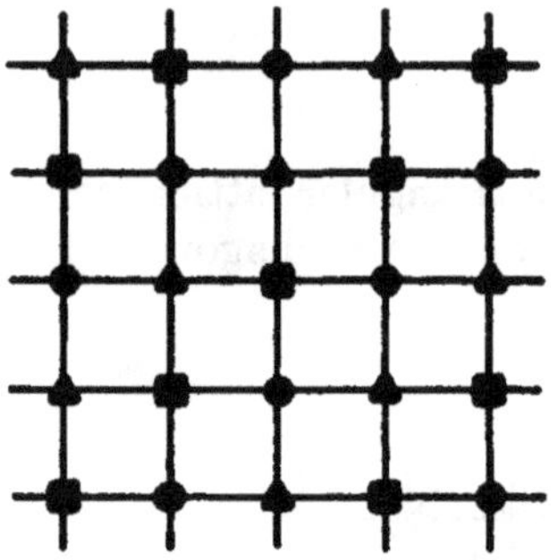

fig. 7

Here every single triangle consists of three sites belonging to the three different sublattices (squares, circles and triangles); reversing all spins of any two of the sublattices the hamiltonian is left unchanged. This can be done in three different ways; so there are three sublattice spin reversal symmetries.

Eight-vertex and three-spin models look very different because of their different symmetries but they have the same number of ground states.

Infact for the three-spin model there are four ground states:

$\sigma_\blacksquare$	$\sigma_\bullet$	$\sigma_\blacktriangle$
+	+	+
+	−	−
−	+	−
−	−	+

as well as for the eight-vertex model:

$(\sigma_i\sigma_j)_J$		$(\sigma_i\sigma_j)_{J'}$	
+	+	+	+
+	+	−	−
−	−	+	+
−	−	−	−

It turns out that one can actually transform the three-spin model into a special case of the eight-vertex model; that transformation changes most of the spins but leaves all the spins within a certain diagonal row unchanged (which also means that the order parameter of the models are the same).

So the three-spin model is a special case of the eight-vertex model in the sense that they have the same spontaneous magnetization, the same free energy and the same spin correlation within a given diagonal row.

Hard hexagons model

Again we have a triangular lattice, on which we put a lattice gas: so instead of spins on sites we have particles on sites. No two particles can sit on the same site nor they can sit on adjacent sites. It's called hard hexagons because gluing the six triangules surrounding a particle they form a hexagon and no two such hexagons can overlap (they're "hard").

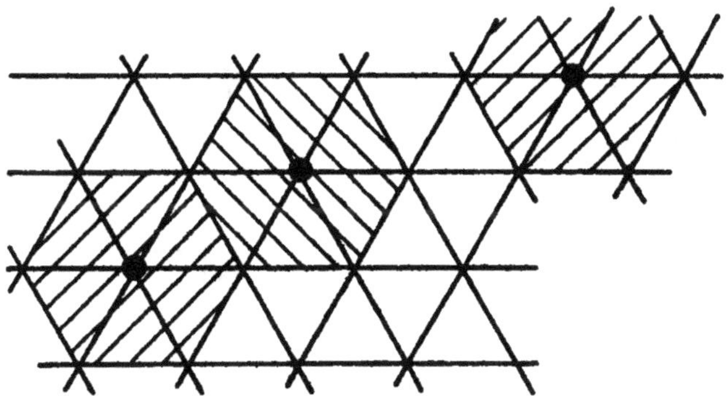

fig. 8

We want to calculate the number of ways of putting n particles on N sites, $g(n, N)$, the maximum number of particles we can put on such a lattice being of course $N/3$. So the partition function is:

$$Z_N = \sum_{n=0}^{N/3} z^n g(n, N)$$

where z is the activity.

Hard hexagons model can be thought as an interaction round a face model (fig. 9).

First let's go from this grand-canonical way of talking about position of sites to the canonical way where we talk about sites. This can be done associating to each site (i) an occupation number

$$\sigma_i = \begin{cases} 0 & \text{if site (i) is empty} \\ 1 & \text{if site (i) is full} \end{cases}$$

The total number of particles is then just the sum of the occupation numbers and the partition function becomes:

$$\begin{aligned} Z_N &= \sum_{\sigma_i} \cdots \sum_{\sigma_N} z^{\sigma_1 + \cdots + \sigma_N} \prod_{edges} (1 - \sigma_i \sigma_j) = \\ &= \sum_{\sigma_i} \cdots \sum_{\sigma_N} \prod_{faces} w(\sigma_i, \sigma_j, \sigma_k, \sigma_l) \end{aligned} \tag{1.13}$$

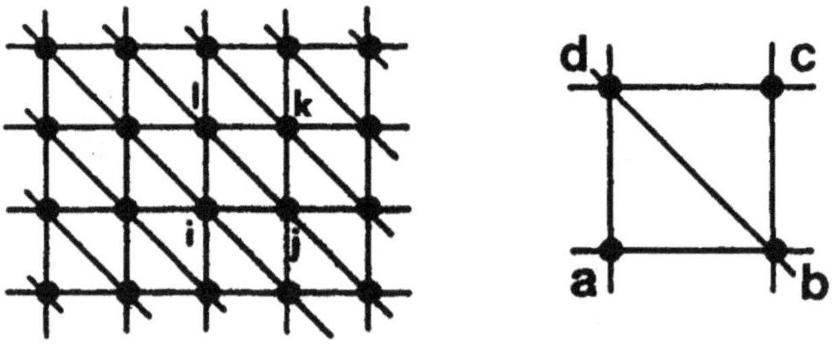

fig. 9

In (1.13) we see that the contribution to Z_N is zero if any two occupation numbers on a triangular face are zero. Besides $w(a, b, c, d)$ clearly is:

$$w(a, b, c, d) = z^{(a+b+c+d)/n}(1 - ab)(1 - bc)(1 - cd)(1 - da)(1 - db)(1 - ca) \tag{1.14}$$

The $(\frac{1}{n})$ in the exponent of z is necessary as site (a), for example, belongs to four different squares.

Table 3 is a list of all these exactly solved IRF models.

Model	Value of a, b, c, d	$\epsilon(a, b, c, d)$
n.n. Ising	$+1, -1$	$-\frac{1}{2}J(ab + cd) - \frac{1}{2}J'(bc + ad)$
n.n.n. Ising	$+1, -1$	$-Jac - J'bd$
8-Vertex	$+1, -1$	$-Jac - J'bd - J_4 abcd$
6-Vertex	$+1, -1$ $+1, -1$	$-Jac - J'bd - J_4 abcd$ with $-J, J', -J_4 \to +\infty$, differences finite
3-Spin	$+1, -1$	$-J(abd + bcd)$
Hard Hexagons	$0, -1$	$-K_B T \log\{z^{\frac{a+b+c+d}{4}}(1 - ab) \cdot$ $\cdot(1 - bc)(1 - cd)(1 - da)(1 - bd)\}$

table 3

Really those are not all the exactly solved IRF models but they are the most significant. For example a six-vertex model with an external electric field has been solved: the solution leads to a linear integral equation.

ORDER PARAMETER

Two models are equivalent if they have the same free energy and the same order parameter. For the Ising model the order parameter is the spontaneous magnetization M_0, the zero denoting that it's evaluated for zero applied magnetic field. There are various way of defining it.

One way is to consider the Ising model in a field H, with a certain magnetization $M(N,H)$; first one takes the thermodynamic limit of $M(N,H)$ and then the limit for $H \to 0^+$:

$$M_0 = \lim_{H\to 0^+} \lim_{N\to\infty} M(N,H)$$

However we shall use the following definition. Take a finite square lattice and fix all the boundary spins to be up (fig. 10):

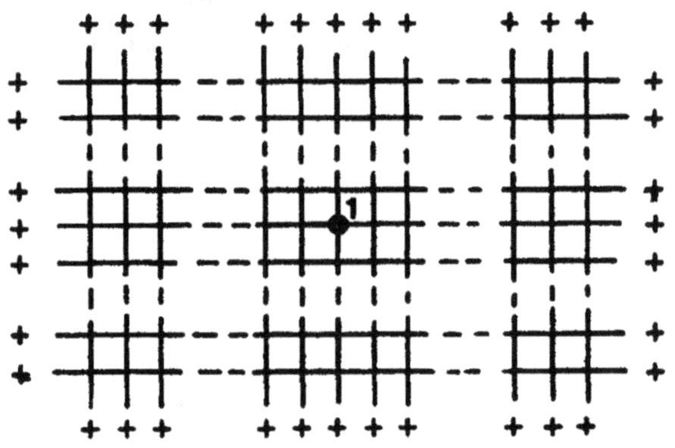

fig. 10

Now look at the center site on some site deep in the middle of the lattice, say (1). The expectation value for σ_1 is then:

$$< \sigma_1 >= \frac{\sum_{conf} \sigma_1 e^{-\beta H}}{\sum_{conf} e^{-\beta H}} \tag{2.1}$$

The hamiltonian of the Ising model (as that of the eight-vertex model) is unchanged by reversing all the spins, so if there are free boundary conditions or cyclic boundary conditions then $< \sigma_1 >$ is zero. Infact to every term in the numerator sum corresponds, by reversing all spins, the same term with opposite sign as σ_1 changes sign while the exponential factor remains unchanged. But we have not given cyclic boundary conditions and it's certainly true that for ferromagnetic systems there are states for which these averages are positive.

This for finite lattices. Now let's consider the case when N, the number of sites of the lattice, goes to infinite. It may be argued that, as the boundary goes away from σ_1, boundary effects shouldn't matter so we should obtain the same result as if we have cyclic boundary conditions and $< \sigma_1 >$ has to decrease as $N \to \infty$. This is infact true for sufficiently high temperatures while for T less than a certain temperature T_c that's no more true and the system in σ_1 remembers the boundary conditions (long range ordered system): $< \sigma_1 >$ as $N \to \infty$ tends to a positive value that we call spontaneous magnetization (fig. 11):

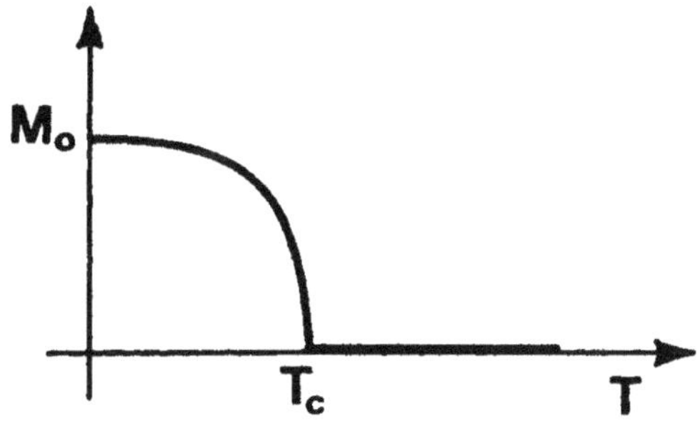

fig. 11

The critical exponent β tell us the way in which M_0 vanishes at T_c. For the Ising model β is $\frac{1}{8}$ so that curve looks almost like a step function.

That's one possible definition of the order parameter of the Ising and the eight-vertex model.

For three-spin and hard hexagon models one uses a different definition. In every case it's useful to look for something that would be zero if there were cyclic boundary conditions in a finite system but is expected not to be zero if one perturbs the boundary conditions.

In the ordered states of both three-spin and hard hexagon models there is a breaking of translational invariance as, depending on the boundary, the particles prefer one of the three sublattices of the triangular lattice; the order parameter then is an expression of that. We could calculate the expectation value of σ_1 (a spin in the middle of the lattice), which will now be the local density of particles at site one;

setting ρ_i as the density on sublattice (i) $(i = 1,2,3)$, we find that for sufficiently high activity $(z > z_c)$ ρ_1 depends on the sublattice on which σ_1 is. If sublattice (1) for example is favoured at the boundary, for $z > z_c$ then $\rho_1 > \rho_2, \rho_3$ while $\rho_2 = \rho_3$. So one can naturally define the order parameter R to be:

$$R = \rho_1 - \rho_2 \tag{2.2}$$

We can plot R as a function of z (fig. 12):

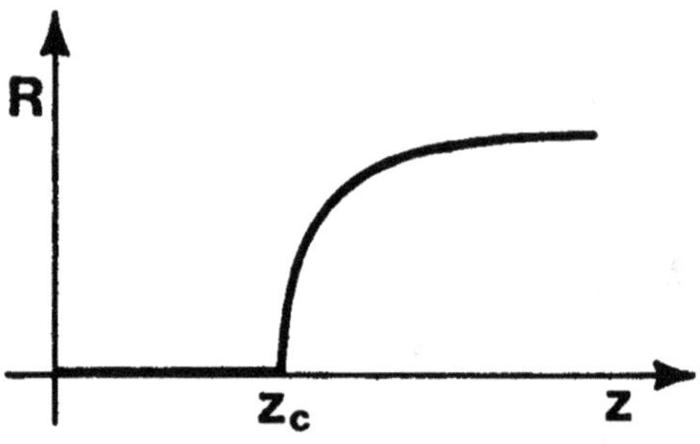

fig. 12

MATHEMATICAL METHODS

In table 4 we illustrate how these models have been solved.

Model	Pfaffians	Fermion Algebra	Bethe Ansatz	Commuting Transfer Matrices	Matrix Inversion Trick	Corner Transfer Matrix
Ising	$\oplus$	$\boldsymbol{\oplus}\boldsymbol{\otimes}$	$\oplus$	$\oplus$	$\oplus$	$\otimes$
Dimer	$\boldsymbol{\oplus}$	$\oplus$	$\oplus$			
6-Vertex			$\boldsymbol{\oplus}\boldsymbol{\otimes}$	$\oplus$	$\oplus$	$\otimes$
8-Vertex			$\oplus$	$\boldsymbol{\oplus}$	$\oplus$	$\boldsymbol{\otimes}$
3-Spin			$\boldsymbol{\oplus}$			$\otimes$
Hard Hexagons					$\boldsymbol{\oplus}$	$\boldsymbol{\otimes}$

$\boldsymbol{\oplus}$: First method used for free energy
$\oplus$: Later method used for free energy
$\boldsymbol{\otimes}$: First method used for order parameter
$\otimes$: Later method used for order parameter

table 4

As there is a simple trick for getting the free energy of all these models, the ***matrix inversion*** trick, so there is a simple trick for getting the order parameter, the ***corner transfer matrix*** trick.

In these lectures we'll talk about the last three method listed in table 4.

Matrix inversion trick for calculating free energy

We have to introduce the transfer matrix. There are many ways of doing it as there are many ways of drawing a lattice. We choose to draw it diagonally, on a cylinder or on a thorus (fig. 13).

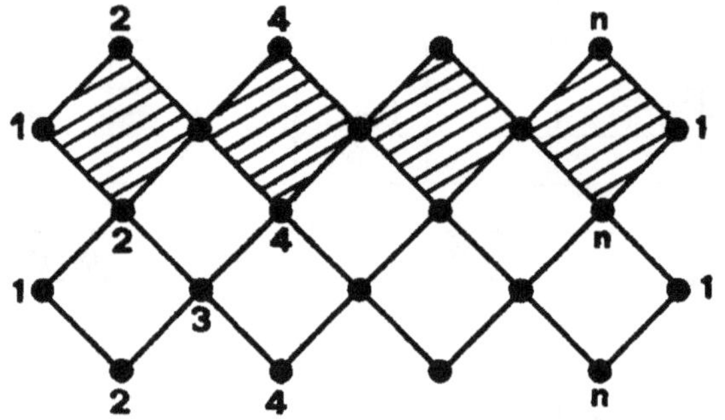

fig. 13

Let's look at a particular row of faces on the lattice, say those one shaded in fig. 13; now look at the spins on the upper and lower segments (fig. 14):

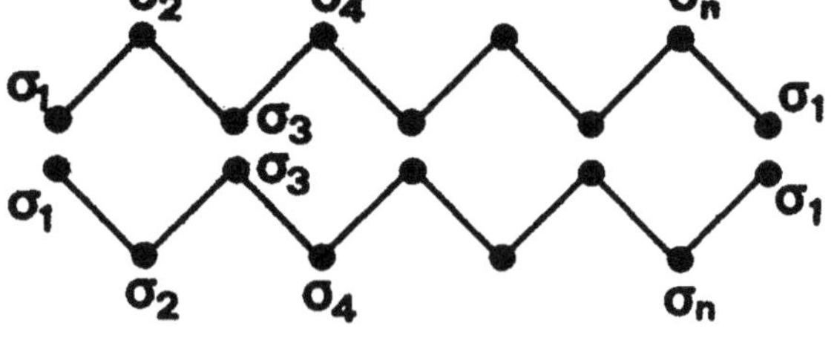

fig. 14

The Boltzmann weight of that row of faces is just the product of the Boltzmann weights of the faces:

$$V(\sigma_1, \ldots, \sigma_n \mid \sigma'_1, \ldots, \sigma'_n) = $$
$$= \prod_{j=1}^{n/2} \{w(\sigma_{2j-1}, \sigma_{2j}, \sigma_{2j+1}, \sigma'_{2j})\delta(\sigma_{2j-1}, \sigma'_{2j-1})\} \tag{3.1}$$

Let's consider series of such rows (fig. 15, a e b)

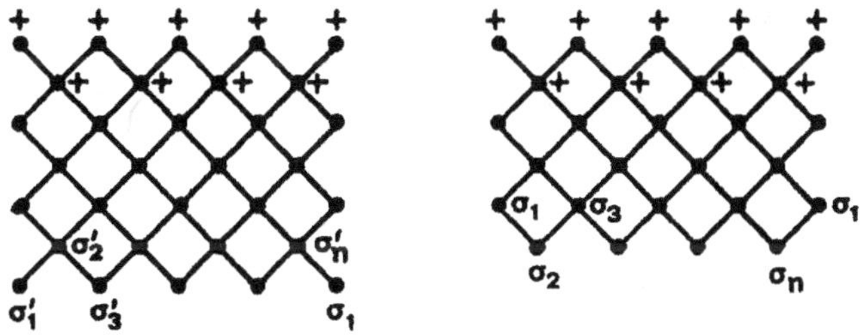

fig. 15 (a) fig. 15 (b)

Here the partition function is:

$$Z = \sum_{\substack{interior \\ spins}} \prod_{faces} w(\sigma_i, \sigma_j, \sigma_k, \sigma_l) \tag{3.2}$$

$$Z = \begin{cases} X(\sigma'_1, \ldots, \sigma'_n) & \text{for fig. (a)} \\ Y(\sigma_1, \ldots, \sigma_n) & \text{for fig. (b)} \end{cases}$$

There is of course a relationship between X and Y:

$$Y(\sigma_1, \ldots, \sigma_n) = \sum_{\sigma'_1 \ldots \sigma'_n} V(\sigma_1 \cdots \sigma_n \mid \sigma'_n \cdots \sigma'_n) X(\sigma'_1 \cdots \sigma'_n) \tag{3.3}$$

That equation may be written as a matrix equation. First define:

$$\phi = \{\sigma_1, \ldots, \sigma_n\} \qquad \phi' = \{\sigma'_1, \ldots, \sigma'_n\}$$

so (3.3) becomes:

$$Y(\phi) = \sum_{\phi'} V(\phi \mid \phi') X(\phi')$$

Now set:

$$\begin{array}{lll}\mathbf{X}= & \text{vector with elements} & X(\phi)\\ \mathbf{Y}= & \text{vector with elements} & Y(\phi)\\ \mathbf{V}= & \text{matrix with elements} & V(\phi \mid \phi')\end{array}$$

Then we obtain for (3.3):

$$\mathbf{Y}=\mathbf{VX} \tag{3.4}$$

The matrix **V** is the diagonal transfer matrix. It has $2^n \times 2^n$ elements because each σ in all these models has two possible values so each ϕ has 2^N possible values. When we add the next row we use almost the same procedure but now everything is shifted of one column and the row transfer matrix becomes:

$$W(\sigma_1,\ldots,\sigma_n \mid \sigma'_1,\ldots,\sigma'_n)=V(\sigma_2,\ldots,\sigma_n,\sigma_1 \mid \sigma'_2,\ldots,\sigma'_n,\sigma'_1)$$

$$\text{and} \tag{3.5}$$

$$\mathbf{Z}=\mathbf{WY}=\mathbf{WVX}$$

where $Z(\sigma_1,\ldots,\sigma_n)$ is the partition function with one more row added. If we now keep up multiplying alternatively by **V** and **W** we can build the whole lattice; fixing top and bottom rows to have values $(+1)$, the partition function turns up to be:

$$Z_N=<+\mid \mathbf{VWVW}\cdots\mathbf{VW}\mid +> \tag{3.6}$$

or, using cyclic boundary conditions:

$$Z_N=\mathrm{Tr}(\mathbf{VWVW}\cdots\mathbf{VW})=\mathrm{Tr}(\mathbf{VW})^m \tag{3.7}$$

where $2m$ is the number of rows, n the numbers of columns and $N=mn$ the number of sites.

If we introduce the eigenvalues of **VW**, say $\lambda_1,\ldots,\lambda_{2^n}$, (3.7) becomes:

$$Z_N=\sum_{r=1}^{2^n}(\lambda_r)^m \tag{3.8}$$

In statistical mechanics one usually works with large systems; it means that one has to consider the limit as m,n tends to infinite. Let's first consider the limit $m\to\infty$, keeping n fixed; the sum in (3.8) will be finite and it will be dominated by the biggest eigenvalue λ_0:

$$m \to \infty \qquad Z_N \sim (\lambda_0)^m$$

Note that for m large but finite the maximum eigenvalue is effectively unique, according to Frobenius theorem for matrices with only positive entries (as **V** and **W**)[19].

In the thermodynamic limit, as the partition function becomes infinite, one wants to know the partition function per site, which is the Nth root of Z_N:

$$k = \exp\left\{-\frac{f}{K_B T}\right\} = (Z_N)^{\frac{1}{N}} = \lambda_0^{\frac{1}{n}} \tag{3.9}$$

So we are interested in calculating the maximum eigenvalue of **VW**. **V** and **W** can both be expressed in terms of the face transfer matrices $\mathbf{U}_i$, whose elements are:

$$\begin{aligned} \mathbf{U}_i(\phi \mid \phi') = \delta(\sigma_1, \sigma_1') \cdots \delta(\sigma_{i-1}, \sigma_{i-1}') \cdot \\ \cdot\, w(\sigma_i \sigma_{i+1} \sigma_i' \sigma_{i-1}) \delta(\sigma_{i+1}, \sigma_{i+1}') \cdots \delta(\sigma_n, \sigma_n') \end{aligned} \tag{3.10}$$

This $\mathbf{U}_i$ definition corresponds to add a single face to the lattice (fig. 16).

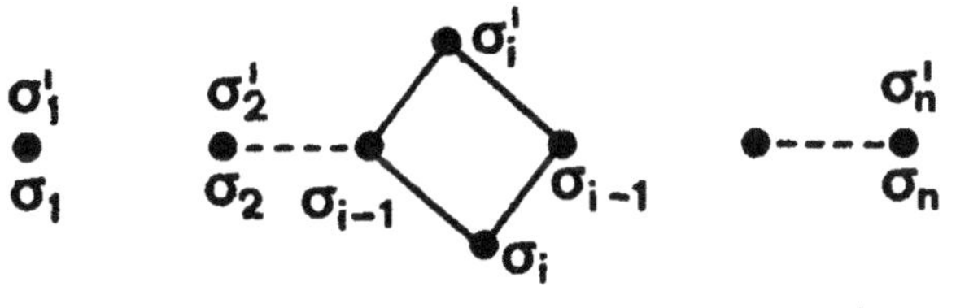

fig. 16

Using (3.10) we have for **V** and **W**:

$$\begin{aligned} \mathbf{V} &= \mathbf{U}_2 \mathbf{U}_4 \cdots \mathbf{U}_n \\ \mathbf{W} &= \mathbf{U}_1 \mathbf{U}_3 \cdots \mathbf{U}_{n-1} \end{aligned} \tag{3.11}$$

Note that: $\mathbf{U}_i \mathbf{U}_j = \mathbf{U}_j \mathbf{U}_i$ if $|i-j| \geq 2$. $\mathbf{U}_i$ for each row has only two non zero entries, i.e. two values for the set ϕ which contributes; infact σ_l for $l \neq i$ must be equal to σ_l' and for $l = i$ can assume the values (± 1).

Clearly the same is true for each column. Then we can rearrange these rows and columns to put $\mathbf{U}_i$ in a block diagonal form with 2×2 blocks. In this form it's particularly easy to invert $\mathbf{U}_i$, as one has only to invert 2×2 matrices; the inverse matrix $\overline{\mathbf{U}}_i$ turns out to be of the same form as $\mathbf{U}_i$. The only difference is in the weight function $\overline{w}$; from the equation

$$\mathbf{U}_i \overline{\mathbf{U}}_i = \mathbf{I}$$

the relation between w and $\overline{w}$ must be:

$$\sum_{d'} w(a,b,c,d')\overline{w}(a,d',c,d) = \delta(b,d) \tag{3.12}$$

for all a, b, c and d.

The transformation $w \to \overline{w}$ inverts $\mathbf{V}$ and $\mathbf{W}$ and generates a similar transformation for the product $\mathbf{VW}$; so it inverts the eigenvalues $\lambda_1, \ldots, \lambda_{2^N}$, taking the biggest to the smallest. But we need some definition which fixes the attention on a particular eigenvalue. So let's define λ_0 as the eigenvalue of $\mathbf{VW}$ with eigenvector corresponding to all positive entries; inverting $\mathbf{VW}$ inverts the eigenvalues but doesn't change the eigenvectors. Then this definition of λ_0 is the same as before for the weight function w while the value of λ_0 is inverted when we replace w with $\overline{w}$.

Using (3.9) we obtain for the free energy the inversion relation:

$$k(w)k(\overline{w}) = 1 \tag{3.13}$$

We now need two statements that for solved models appear to be true:

(i) in some sense $k(\overline{w})$ is the analytic continuation of $k(w)$;

(ii) if one rotates the lattice of 90 degrees the partition function and then $k(w)$ don't change.

The property of symmetry (ii) with the inversion relation (3.13)and the property of analyticity (i) are enough to determine the free energy of the system.

As an illustration let's look at the zero-field Ising model. From (1.9):

$$w(a,b,c,d) = \exp(\mathsf{K}ac + \mathsf{K}'bd)$$

$$\mathsf{K} = \frac{J}{K_B T} \qquad \mathsf{K}' = \frac{J'}{K_B T}$$

Now using (3.12):

$$\overline{w}(a,b,c,d) = \frac{-i}{2\sinh 2\mathsf{K}'} e^{-\mathsf{K}ac + (\mathsf{K}' + i\frac{\pi}{2})bd} \tag{3.14}$$

It is a real number because of the factor $e^{i\frac{\pi}{2}bd}$ that's just $+1$ or -1.

Multiplying w by α the partition function is multiplied by α^N; this corresponds to simply multiplying the partition function per site by α. We obtain:

$$k(\alpha w) = \alpha k(w)$$

Now use this relation and the inversion ones (3.13) and (3.14):

$$k(\mathsf{K},\mathsf{K}')k(-\mathsf{K},\mathsf{K}+i\frac{\pi}{2}) = 2i\sinh 2\mathsf{K}' \tag{3.15}$$

Besides the symmetry relation (ii) for our model is:

$$k(\mathsf{K},\mathsf{K}') = k(\mathsf{K}',\mathsf{K}) \tag{3.16}$$

As we want to do a low temperature expansion it's useful to introduce the variables u and u' defined by:

$$u = e^{-4\mathsf{K}} \qquad u' = e^{-4\mathsf{K}'}$$

and the function $\Lambda(u,u')$ defined by:

$$k(\mathsf{K},\mathsf{K}') = e^{\mathsf{K}+\mathsf{K}'}\Lambda(u,u')$$

In the low temperature limit u and u' become zero and Λ is just 1.

Now we can express (3.15), (3.16) in terms of Λ, u and u':

$$\Lambda(u,u')\Lambda(\frac{1}{u},u') = 1 - u' \tag{3.15 bis}$$

$$\Lambda(u,u') = \Lambda(u',u) \tag{3.16 bis}$$

Letting u be arbitrary we'll find a low temperature expansion in powers of u'. This means to perturb the ground state with all spins plus fixing first one arbitrary spin to be minus, then two and so on. Finally:

$$\Lambda(u,u') = 1 + \frac{uu'}{1-u} + \frac{uu'^2}{(1-u)^3} + \cdots \tag{3.17a}$$

$$\Lambda(\frac{1}{u},u') = 1 - \frac{u'}{1-u} - \frac{u^2u'^2}{(1-u)^3} + \cdots \tag{3.17b}$$

These expansions are true for $u(\frac{1}{u})$ less than 1. By formally replacing in (3.17 a) u by $\frac{1}{u}$ one can see that (3.17 b) is just the analytic continuation of $\Lambda(u,u')$ throught the point $u = 1$, according to the statement (i).

Substituting (3.17) into the left hand side of (3.15 bis) one obtain the correct result to the available order:

$$\Lambda(u,u')\Lambda(\frac{1}{u},u') = 1 - u' + O(u'^2)$$

It wouldn't be difficult, even if one didn't know anything about the Onsager solution, to go on with the expansion (3.17). Let's suppose the following to be the general form of that expansion:

$$\Lambda(u,u') = 1 + \sum_{r=1}^{\infty} \frac{c_r(u)u'^r}{(1-u)^{2r-1}} \tag{3.18}$$

where $c_r(u)$ is a finite degree polynomial in u. (3.18) may be considered as an assumption about the analytic character of the solution. This assumption together with (3.15 bis), (3.16 bis) determines the coefficients $c_1(u), c_2(u), \ldots$, ecc. We'll give an inductive proof of this fact, supposing that $c_1(u), c_2(u), \ldots, c_{r-1}(u)$ are known and determining $c_r(u)$, which has to be a finite polynomial:

$$c_r(u) = \gamma_0 + \gamma_1 u + \cdots + \gamma_{r-1}u^{r-1} + \gamma_r u^r + \cdots \tag{3.19}$$

For (3.16 bis) if we know the first $(r-1)$ coefficients in the expansion in powers of u', by symmetry we also know these coefficients in the expansion in powers of u, so that we can know $\gamma_0, \ldots, \gamma_{r-1}$. Then substituting (3.18) for $\Lambda(u,u')$ in the inversion relation (3.15 bis) we obtain by comparing the coefficients of u'^r:

$$\frac{c_r(u)}{(1-u)^{2r-1}} + \frac{c_r(\frac{1}{u})}{(1-\frac{1}{u})^{2r-1}} = \text{known}$$

That multiplied by $(1-u)^{2r-1}$ becomes:

$$c_r(u) - u^{2r-1}c_r(\frac{1}{u}) = \text{known}$$

Using (3.19) we can rewrite this expression and by comparing coefficients with equal powers of u we'll obtain equations for $\gamma_r - \gamma_{r-1}, \gamma_{r+1} - \gamma_{r-2}, \ldots, \gamma_{2r-1} - \gamma_0, \gamma_{2r}, \gamma_{2r+1}, \ldots$; but we know $\gamma_0, \ldots, \gamma_{r-1}$, so we can know $\gamma_r, \gamma_{r+1}, \ldots$. We have then proved that once $c_1(u), \ldots, c_{r-1}(u)$ are given we can calculate $c_r(u)$ for successive r and obtain for the Ising model with zero-field the free energy.

One can think to calculate with the same procedure the free energy for the Ising model in a field. In that case there is another variable, due just to the field H, which enters into the partition function through $h = \frac{H}{K_B T}$. We define u and u' as before and in addition we introduce μ:

$$\mu = e^{-2h}$$

It's now possible to write the free energy per site k using $\Lambda(u, u', \mu)$:

$$k = e^{K+K'+h}\Lambda(u, u', \mu)$$

We still have an inversion relation:

$$\Lambda(u, u', \mu)\Lambda\left(\frac{1}{u}, u', \frac{1}{\mu}\right) = 1 - u' \tag{3.20a}$$

and a symmetry relation:

$$\Lambda(u, u', \mu) = \Lambda(u', u, \mu) \tag{3.20b}$$

Let's look at the low temperature expansion for $\Lambda(u, u', \mu)$, still up to the second order in u':

$$\Lambda(u, u', \mu) =$$

$$1 + \frac{\mu u u'}{1 - \mu u} + \left(\frac{\mu^2 u(1 + \mu u)^2}{(1 - \mu u)^2(1 - \mu^2 u)} - \frac{2\mu^2 u^2 + \mu^3 u^3}{1 - \mu u}\right)u'^2 + O(u'^3)$$

If we now use this expression of $\Lambda(u, u', \mu)$ in (3.20 a), (3.20 b) we no longer have enough informations to determine the coefficients of $u', u'^2, \ldots$.

However it's a good method and perhaps introducing in some way other relations (duality, extrasymmetries, ...) one can think to determine the free energy of a three-dimensional model.

Commuting transfer matrices

The way to get not just the free energy but also the order parameter is to use commuting transfer matrices, and so the eight-vertex model was originally solved by Baxter.

Introduce an ordinary square lattice with m rows and n columns (fig. 17)

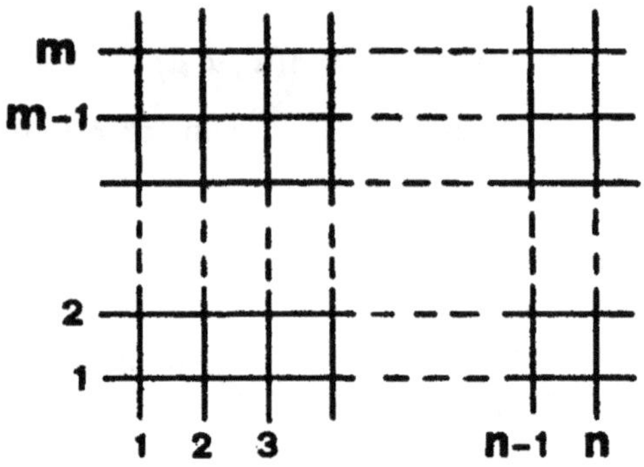

fig. 17

A typical row looks now like in (fig. 18):

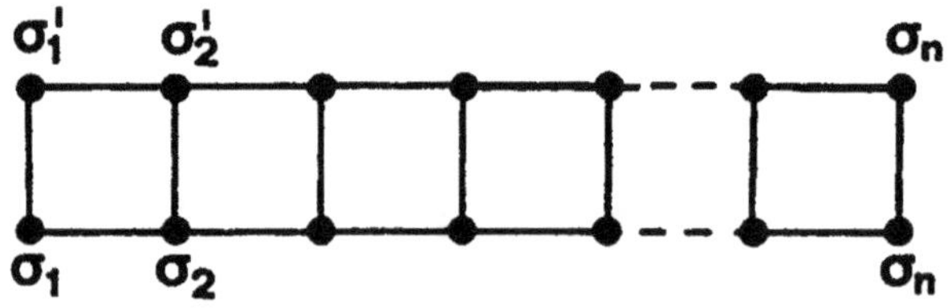

fig. 18

The transfer matrix has elements:

$$V(\sigma_1, \ldots, \sigma_n \mid \sigma_1', \ldots, \sigma_n') = \prod_{j=1}^{n} w(\sigma_j, \sigma_{j+1}, \sigma_{j+1}', \sigma_j') \tag{3.21}$$

As usual set: ϕ_i= all spins on row (i).

Then the partition function is:

$$
\begin{aligned}
Z_N &= \sum_{\phi_1}\sum_{\phi_2}\cdots\sum_{\phi_m} V(\phi_1 \mid \phi_2)V(\phi_3 \mid \phi_3)\cdots V(\phi_m \mid \phi_1) = \\
&= \sum_{\phi_1} (\mathbf{V}^m)_{\phi_1\phi_1} = \mathrm{Tr}\mathbf{V}^m
\end{aligned}
\tag{3.22}
$$

Pictorially this equation means:

$Z_N =$

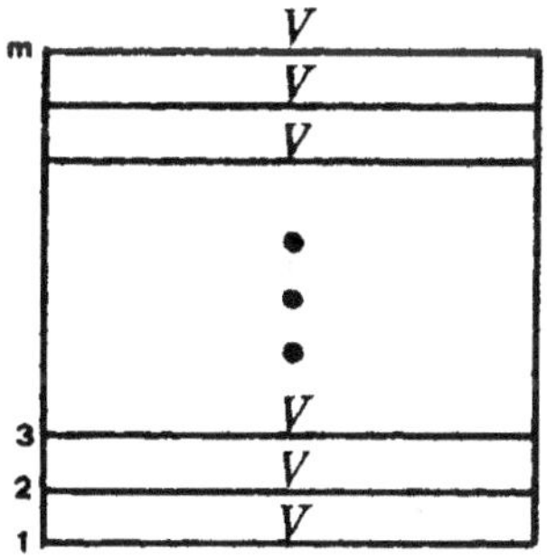

$= \mathrm{Tr}\,\mathbf{V}^m$

As an example let's look again at the Ising model. Set $K = \frac{J}{K_BT}$, $K' = \frac{J'}{K_BT}$. Then:

$$\mathbf{V} = \mathbf{V}(K, K')$$

What Onsager himself noticed is that two transfer matrices $\mathbf{V}(K, K')$ and $\mathbf{V}(L, L')$ commute if the four parameters K, K', L, L' satisfy the relation:

$$\sinh 2K \sinh 2K' = \sinh 2L \sinh 2L' \tag{3.23}$$

We set:

$$X = (\sinh 2K \sinh 2K')^{-1}$$

Equation (3.23) then says that two different transfer matrices with the same value of X commute. We can draw the line X = const in the K,K' plane. Any two matrices on this line will then commute.

Let's now go back to the general IRF model and let's consider two models, one with a weight function w and a transfer matrix $\mathbf{V}$, the other with a different weight function w' and a different transfer matrix $\mathbf{V}'$. An element of the product $\mathbf{V}\mathbf{V}'$ can be expressed as:

$$(\mathbf{VV}')_{\sigma_1\cdots\sigma_n|\sigma_1'\cdots\sigma_n'} = \sum_{\tau_1\cdots\tau_n} V(\sigma_1\cdots\sigma_n \mid \tau_1\cdots\tau_n)V'(\tau_1\cdots\tau_n \mid \sigma_1'\cdots\sigma_n') =$$

$$= \sum_{\tau_1\cdots\tau_n}\prod_{j=1}^{n} w\begin{pmatrix}\tau_j & \tau_{j+1}\\ \sigma_j & \sigma_{j+1}\end{pmatrix} w'\begin{pmatrix}\sigma_j' & \sigma_{j+1}'\\ \tau_j & \tau_{j+1}\end{pmatrix} =$$

(3.24)

$=$

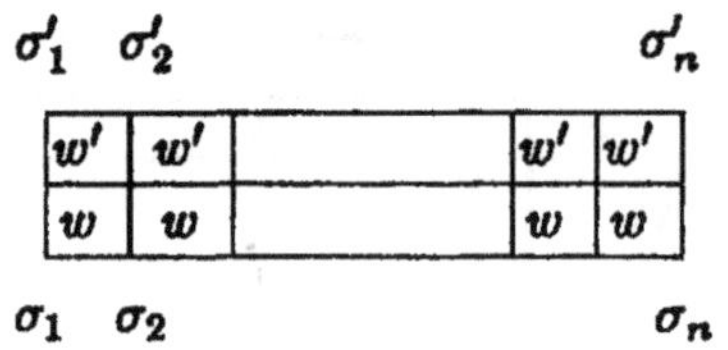

This, as can also be seen graphically, is the partition function of a two row lattice. We may write (3.24) in a simpler way if we suppress the σ_i and σ_i' indices and set:

$$S_j(\tau_j, \tau_{j+1}) = w\begin{pmatrix}\tau_j & \tau_{j+1}\\ \sigma_j & \sigma_{j+1}\end{pmatrix} w\begin{pmatrix}\sigma_j' & \sigma_{j+1}'\\ \tau_j & \tau_{j+1}\end{pmatrix} \tag{3.25}$$

Then:

$$(\mathbf{VV}')_{\sigma_1\cdots\sigma_n|\sigma_1'\cdots\sigma_n'} = \sum_{\tau_1\cdots\tau_n} \mathbf{S}_1(\tau_1,\tau_2)\cdots\mathbf{S}_n(\tau_n,\tau_1) = \mathrm{Tr}\,\mathbf{S}_1\cdots\mathbf{S}_n \tag{3.26}$$

where each $\mathbf{S}_j$ is just a 2×2 matrix with elements $S_j(\tau,\tau')$. Similarly:

$$(\mathbf{V}'\mathbf{V})_{\sigma_1\cdots\sigma_n|\sigma_1'\cdots\sigma_n'} = \mathrm{Tr}\,\mathbf{S}_1'\cdots\mathbf{S}_n'$$

where $\mathbf{S}_j'$ is $\mathbf{S}_j$ with w and w' interchanged.

We want $\mathbf{V}$ and $\mathbf{V}'$ to commute, i.e.:

$$\mathbf{VV}' = \mathbf{V}'\mathbf{V}$$

A sufficient condition is the existence of n 2×2 matrices $\mathbf{M}_1,\ldots,\mathbf{M}_n$ such that:

$$\mathbf{S}_j = \mathbf{M}_j\mathbf{S}_j'\mathbf{M}_{j+1}^{-1} \qquad j = 1,\ldots,n$$

i.e.

$$\mathbf{S}_j\mathbf{M}_{j+1} = \mathbf{M}_j\mathbf{S}_j' \tag{3.27}$$

$\mathbf{S}_j$ is a function of σ_j, σ_{j+1}, σ'_j, σ'_{j+1} while for satisfying (3.27) it is sufficient that $\mathbf{M}_j$ is a function of σ_j and σ'_j.

So define:

$$(\mathbf{M}_j)_{\tau\tau'} = w''\begin{pmatrix} \sigma_j & \tau_j \\ \tau'_j & \sigma'_j \end{pmatrix}$$

This last definition together with the one (3.25) of the elements of $\mathbf{S}_j$ allows us to write down explicitly (3.27) as:

$$\sum_{\tau''} w\begin{pmatrix} \tau & \tau'' \\ \sigma_j & \sigma_{j+1} \end{pmatrix} w'\begin{pmatrix} \sigma'_j & \sigma'_{j+1} \\ \tau & \tau'' \end{pmatrix} w''\begin{pmatrix} \sigma_{j+1} & \tau'' \\ \tau' & \sigma'_{j+1} \end{pmatrix} =$$

$$= \sum_{\tau''} w''\begin{pmatrix} \sigma_j & \tau \\ \tau'' & \sigma'_j \end{pmatrix} w'\begin{pmatrix} \tau'' & \tau' \\ \sigma_j . & \sigma_{j+1} \end{pmatrix} w\begin{pmatrix} \sigma'_j & \sigma'_{j+1} \\ \tau'' & \tau' \end{pmatrix} \qquad (3.28)$$

Let's simplify the notation. Set for the external spins:

$$\tau, \sigma_j, \sigma_{j+1}, \tau', \sigma'_{j+1}, \sigma'_j = a, b, c, d, e, f$$

and for the internal spin:

$$\tau'' = g$$

Besides define the rotated form w'_r of w':

$$w'_r\begin{pmatrix} d & c \\ a & b \end{pmatrix} = w'\begin{pmatrix} a & d \\ b & c \end{pmatrix}$$

When we do that (3.28) becomes:

$$\sum_g w\begin{pmatrix} a & g \\ b & c \end{pmatrix} w'_r\begin{pmatrix} e & g \\ f & a \end{pmatrix} w''\begin{pmatrix} c & g \\ d & e \end{pmatrix} = \sum_g w''\begin{pmatrix} b & a \\ g & f \end{pmatrix} w'_r\begin{pmatrix} d & c \\ g & b \end{pmatrix} w\begin{pmatrix} f & e \\ g & d \end{pmatrix} \qquad (3.29)$$

Graphically it means:

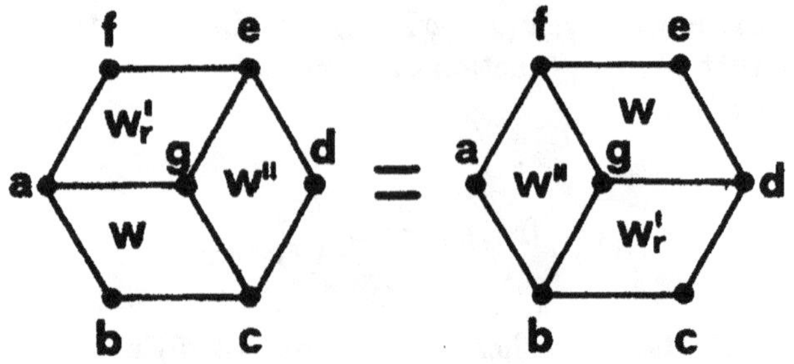

These equations have to be true for all values of the six external spins, so there are 2^6 $(= 64)$ equations to be satisfied.

For the Ising model these equations reproduce the star triangle relations and the Onsager's commutation relations. Infact for this model, as we've seen (1.9):

$$w\begin{pmatrix} d & c \\ a & b \end{pmatrix} = e^{Kac+K'bd}$$

and similarly for w' and w''; so (3.29) reads:

$$e^{K'ca+L'ae+M'ec}\cosh(Kb + Lf + Md) =$$
$$= e^{K'fd+L'bd+m'bf}\cosh(Kl + Lc + Ma)$$

and these 64 equations factor into two sets of 8 equations:

$$2\cosh(Kl + Lc + Ma) = Ae^{K'ca+L'ae+M'ec}$$
$$2\cosh(Kb + Lf + Md) = Ae^{K'fd+L'bd+M'bf}$$

Graphically:

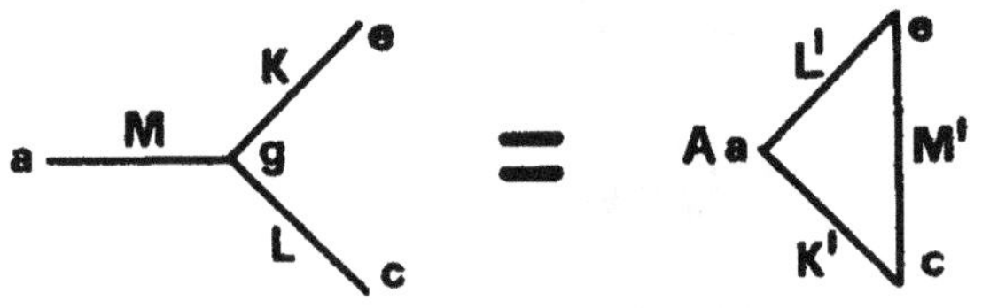

These are just the ***star triangle*** relations.

We are looking for a solution of (3.29), w being given. A trivial one is:

$$w' = w\,; \qquad w''\begin{pmatrix} d & c \\ a & b \end{pmatrix} = \delta(a,c)\delta(b,d)$$

but we're not interested in it as this solution only implies that **V** commutes with itself.

Anyway we know the form of w for the eight-vertex model (see (1.10)):

$$\begin{aligned} \epsilon(a,b,c,d) &= -Jac - J'bd - J_4 abcd \\ w(a,b,c,d) &= Re^{Kac+Lbd+Mabcd} \end{aligned} \tag{3.30}$$

Clearly for w'_r and w'' we can suppose the same form:

$$\begin{aligned} w'_r(a,b,c,d) &= R' e^{K'ac+L'bd+M'abcd} \\ w''(a,b,c,d) &= R'' e^{K''ac+L''bd+M''abcd} \end{aligned}$$

Star triangle relation are now:

$$\begin{aligned} &e^{Lca+L'ac+L''ec} \cosh(Kb + K'f + K''d + Mcba + M'afe + M''edc) = \\ &= e^{Lfd+L'db+L''bf} \cosh(Ke + K'c + K''a + Mfed + M'dcb + M''baf) \end{aligned} \tag{3.31}$$

for all values (± 1) of a, b, c, d, e, f. If K, L, M are given we have only six parameters K', L', M', K'', L'', M'' at our disposal for solving all these 64 equations! Fortunately there are many simplifications because of the symmetry properties of (3.31). Infact negating a, c, e leaves the equations unchanged; the same is true negating b, d and f. So we're left with 16 equations. Further interchanging a with d, b with e and c with f just changes the two sides of (3.31) and the 16 equations reduce to six.

Now choose $M = M' = M''$ and the six equations reduce to three:

$$\begin{aligned} e^{2L'+2L''} &= \frac{\cosh(K + K' + K'' - M)}{\cosh(-K + K' + K'' + M)} \\ e^{2L''+2L} &= \frac{\cosh(K + K' + K'' - M)}{\cosh(K - K' + K'' + M)} \\ e^{2L+2L'} &= \frac{\cosh(K + K' + K'' - M)}{\cosh(K + K' - K'' + M)} \end{aligned} \tag{3.32}$$

If we introduce $\Delta(K, L, M)$,

$$\Delta(K, L, M) = -\sinh 2K \sinh 2L - \tanh 2M \cosh 2K \cosh 2L \tag{3.33}$$

Then eliminating in (3.32) K'', L'' we find:

$$\Delta(K, L, M) = \Delta(K', L', M)$$

This means that two transfer matrices commute if they have the same value of M and the same value of Δ. But we have to fix four parameter, a, b, c, d, so, apart from a normalization factor, we are left with one degree of freedom which permits to construct a class of transfer matrices commuting with each other.

Now we want to parametrize a, b, c, d, and so w, in terms of elliptic functions, i.e. functions of two variables: the *nome* q and the argument u. Usually q is regarded as a real constant, $0 < q < 1$, while u is a complex number. We can also introduce the modulus κ, related to q

$$\kappa = 4q^{\frac{1}{2}} \prod_{n=1}^{\infty} \left(\frac{1 + q^{2n}}{1 + q^{2n-1}} \right)^4$$

and set:

$$\frac{1-\kappa^2}{\kappa^2} = (\Delta^2 - 1) \cosh(2M) \tag{3.34}$$

Note that so we can distinguish the two cases $\kappa^2 > 1$, $\kappa^2 < 1$:

$\lvert\Delta\rvert < 1$	disordered phase
$\lvert\Delta\rvert > 1$	ordered phase

with transition at $|\Delta| = 1$. As a special case when $K, L, M > 0$ and $\Delta < -1$ we have the ferromagnetic ordered phase, with $0 < \kappa < 1$.

Now define:

$$\operatorname{snh}(u, q) = 2q^{\frac{1}{4}} \sinh u \prod_{n=1}^{\infty} \frac{(1 - q^{2n}e^{2u})(1 - q^{2n}e^{-2u})}{(1 - q^{2n-1}e^{2u})(1 - q^{2n-1}e^{-2u})} = \operatorname{snh}(u) \tag{3.35}$$

This is a double periodic function (fig. 19), in the sense that:

$$\operatorname{snh}(u + \pi i) = -\operatorname{snh}(u)$$
$$\operatorname{snh}(u + 2\tau) = \operatorname{snh}(u); \qquad q = e^{-2\tau}$$

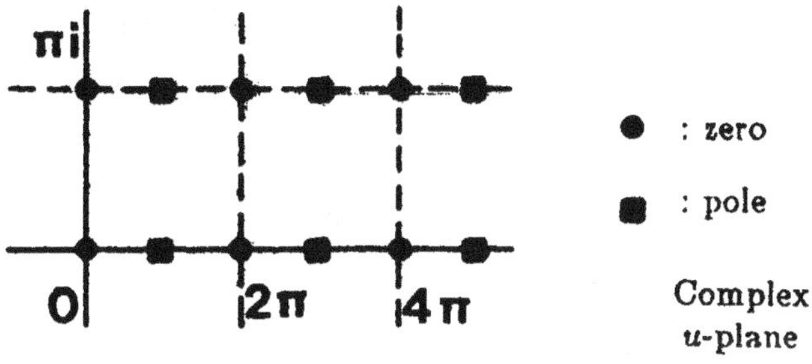

fig. 19

If one knows such a function within a period rectangle, one knows it at any point in the complex plane.

If we parametrize:

$$e^{-2K} = \operatorname{snh}(u, q) \tag{3.36a}$$

$$e^{-2L} = \operatorname{snh}(\lambda - u, q) \tag{3.36b}$$

$$e^{2M} = \operatorname{snh}(\lambda, q) \tag{3.36c}$$

then Δ, given in (3.33), turns out to be a function of only λ, q.

In terms of these variables the ordered ferromagnetic phase discussed above corresponds to the assumption:

$$0 < u < \lambda < r\ .$$

Now regarding q, λ as fixed and u as a variable the transfer matrix is:

$$\mathbf{V} \equiv \mathbf{V}(u)$$

and as M, λ are also fixed we obtain:

$$\mathbf{V}(u)\mathbf{V}(v) = \mathbf{V}(v)\mathbf{V}(u)$$

for all complex numbers u and v.

Let's look at the inverse weight function defined by (3.12). As w has the form (3.30), using the inversion and symmetry relations:

$$k(w)k(\overline{w}) = 1$$
$$k(w) = k(w_r)$$

we may write $\overline{w}$ as:

$$\overline{w} = \overline{R}e^{\overline{K}ac+\overline{L}bd+\overline{M}abcd} \tag{3.37}$$

where

$$\overline{K} = K + i\frac{\pi}{2}, \qquad \overline{M} = M,$$

$$e^{2\overline{L}} = e^{-2L}\frac{\sinh(2K-2M)}{\sinh(2K+2M)},$$

$$\overline{R} = \frac{-i}{R\sqrt{4\sinh(2K-2M)\sinh(2K+2M)}}$$

with these relations we can easily verify that:

$$\overline{\Delta} = \Delta(\overline{K},\overline{L},\overline{M}) = \Delta(K,L,M) = \Delta \tag{3.38}$$

But Δ is a function of q, λ only, so (3.38) means that: $\overline{q},\overline{\lambda} = q,\lambda$.

Besides for (3.36 a) and for the definition (3.37) of $\overline{K}$ it turns out that:

$$\overline{u} = -u$$

i.e. negating u takes w to $\overline{w}$!

We can now demonstrate that all weight functions are entire functions of u.

First choose κ in (3.34) so that:

$$w(+,+,+,+) = Re^{K+L+M} = \rho H(\lambda)\Theta(u)\Theta(\lambda-u) \tag{3.39}$$

$$H(u) = 2q^{\frac{1}{4}}\sinh(u)\prod_{n=1}^{\infty}(1-q^{2n}e^{2u})(1-q^{2n}e^{-2u})$$

$$\Theta(u) = \prod_{n=1}^{\infty}(1-q^{2n-1}e^{2u})(1-q^{2n-1}e^{-2u})$$

Then (3.35) becomes:

$$\operatorname{snh}(u) = \frac{H(u)}{\Theta(u)}$$

For $w(+,+,+,-)$ we now obtain:

$$w(+,+,+,-) = Re^{K-L-M} = w(+,+,+,+)e^{-2L-2M} =$$
$$= \rho H(\lambda)\Theta(u)\Theta(\lambda-u)\frac{H(\lambda-u)}{\Theta(\lambda-u)}\frac{\Theta(\lambda)}{H(\lambda)} = \rho\Theta(\lambda)\Theta(u)H(\lambda-u)$$

where we've used the parametrization (3.36 b,c) for e^{-2L-2M}. Similarly, go from $\overline{R}$ to $\overline{\rho}$:

$$\overline{w}(+,+,+,+) = \overline{R}e^{\overline{K}+\overline{L}+\overline{M}} = \overline{\rho}H(\lambda)\Theta(-u)\Theta(\lambda+u)$$

Then:

$$\rho\overline{\rho} = \left(h(\lambda-u)h(\lambda+u)\right)^{-1}$$
$$h(u) = \Theta(0)H(u)\Theta(u)$$

We now rewrite the inversion and symmetry relations together with the periodicity and analyticity properties in terms of this parametrization:

(i) Inversion: $k(\rho,u)k(\overline{\rho},-u) = 1$
Define $k(u) = k(1,u)$, so that $k(\rho,u)$ is:
$k(\rho,u) = \rho k(u)$ and $k(u)k(-u) = \frac{1}{\rho\overline{\rho}} = h(\lambda-u)h(\lambda+u)$

(ii) Symmetry: $k(u) = k(\lambda-u)$; infact interchanging K and L merely replaces u by $\lambda - u$.

(iii) Periodicity: $\ln k(u+\pi i) = \ln k(u)$

(iv) Analiticity: $\ln k(u)$ is analytic in the strip $0 \leq \mathrm{Re}(u) \leq \lambda$ (fig.(20)). One may control this property for example on the first few terms of the low temperature expansion.

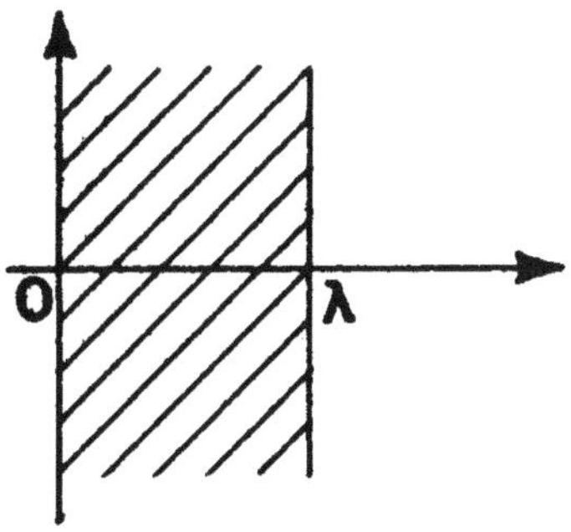

fig. 20

While (i),(ii) and (iii) are rigorous properties, (iv) is like an assumption. Anyway if we accept all of (i)...(iv) we have enough information to determine the free energy $k(u)$.

Infact from (iii) and (iv) we can obtain a Fourier expansion of $\ln k(u)$:

$$\ln k(u) = \sum_{r=-\infty}^{\infty} c_r e^{2ru} \tag{3.40}$$

Then we use this expression in the symmetry relation (ii) and have:

$$c_{-r} = e^{2r\lambda} c_r$$

Finally we take logs of the inversion relation (i) and consider the series (3.40) both for $k(u)$ and for $k(-u)$. Note that so analyticity must apply not only to the right side of the immaginary axis (see fig. 20) or to the axis itself, but has to continue a little bit into the left side, say for a distance ϵ ($\epsilon \ll 1$); once again that is consistent with the low temperature expansion.

However through this procedure we get all the information we need to determine the coefficients c_0, c_1, c_{-1}, ... Then we put them back into (3.40) and obtain the following expression for $\ln k(u)$:

$$\ln k(u) = \ln\left(q^{\frac{1}{4}}\Theta(0)e^{\lambda}\right) - \sum_{r=1}^{\infty} \frac{\left(e^{-2r\lambda} + q^r e^{2r\lambda}\right)\left(e^{2ru} + e^{2r(\lambda-u)}\right)}{r\left(1-q^r\right)\left(1+e^{2r\lambda}\right)} \tag{3.41}$$

This is the solution for the free energy of the eight-vertex model. Similar arguments work for the hard hexagon model: it has the same free energy as a particular eight-vertex model.

The calculation could be done much more rigorously. One has infact to write down the equations for the eigenvalues of the transfer matrix and take the limit as that matrix becomes infinitely large; anyway we get the same results.

Corner transfer matrix

In the star triangle equations (3.29) w, w', w'' with the parametrization (3.36) are functions of q, λ and u, the first two being fixed parameters while u is a variables such that:

$$\begin{aligned} w &= w(u) \\ w' &= w(u') \\ w'' &= w(u'') \\ u'' &= u' - u \end{aligned} \tag{3.42}$$

Later on we'll need this property which is very closely related to the ***transformation to a different kernel*** that is used in the Bethe Ansatz of the six-vertex model solution.

We now introduce the corner transfer matrix, which can be defined for any planar lattice model with finite range interactions; for definiteness we consider an IRF model.

Draw a lattice as that in fig. 21 and associate to each site (i) a spin σ_i which assumes values $+1$ or -1.

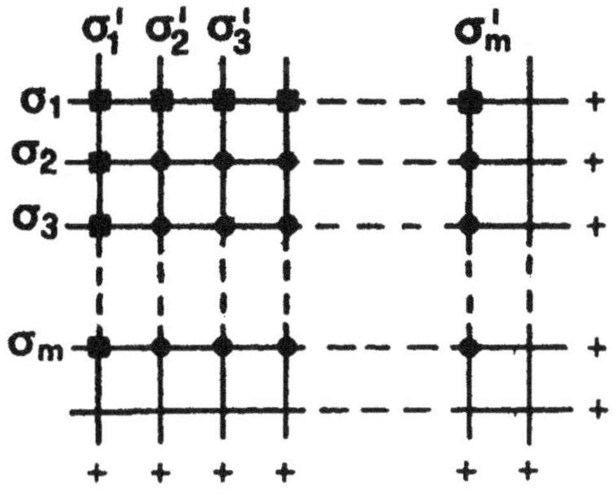

fig. 21

One can think this lattice as the bottom-right corner of a bigger lattice with fixed spin up boundary conditions, according to a possible ground state.

We define the corner transfer matrix as:

$$A(\sigma_1 \cdots \sigma_m \mid \sigma'_1 \cdots \sigma'_m) = \delta(\sigma_1, \sigma'_1) \sum_{\substack{internal\\ spins}} \prod_{faces} w(\sigma_i, \sigma_j, \sigma_k, \sigma_l) \quad (3.43)$$

For a given value of the matrix indices we can think of A as the partition function of the corner lattice; σ_1 and σ'_1 being both the upper left corner spins, the matrix is defined as zero unless $\sigma_1 = \sigma'_1$. Set:

$$\phi = \{\sigma_1, \ldots, \sigma_m\} \qquad \phi' = \{\sigma'_1, \ldots, \sigma'_m\}$$

so A can be written as $A(\phi \mid \phi')$. Now if we consider the full lattice (fig. 22 a) we have to introduce three other corner transfer matrices which are similarly defined (fig. 22 b, c, d).

$A(\phi \mid \phi')$

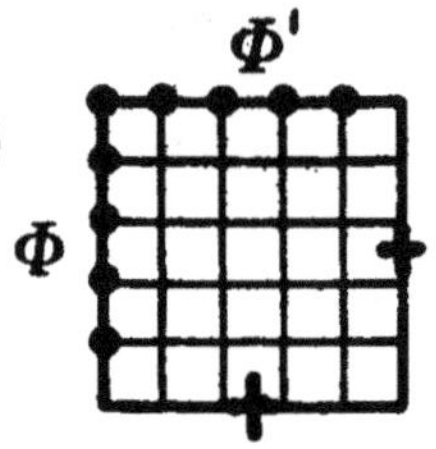

fig. 22 (a)

$B(\phi' \mid \phi'')$

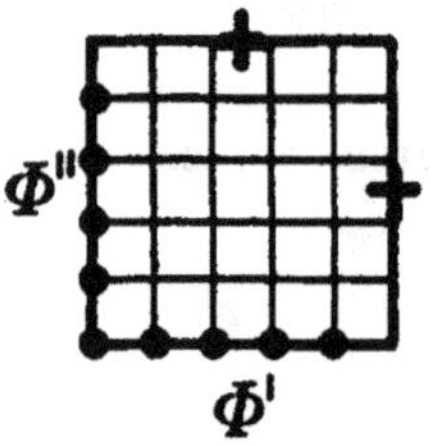

fig. 22 (b)

$D(\phi''' \mid \phi)$

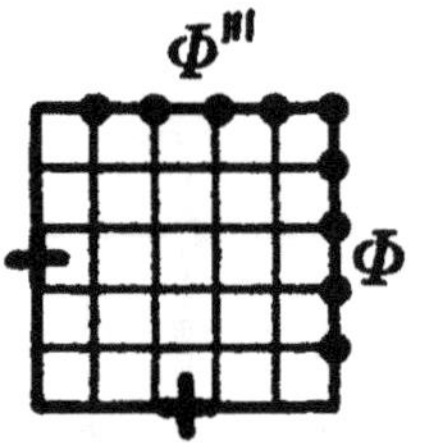

fig. 22 (c)

$C(\phi'' \mid \phi''')$

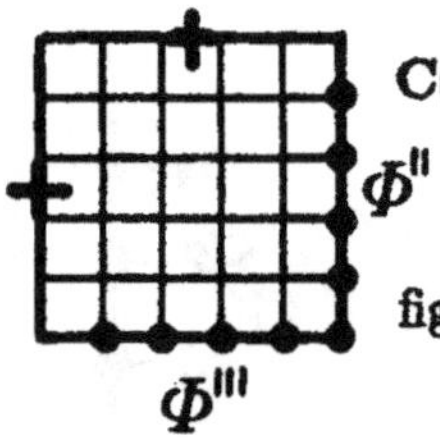

fig. 22 (d)

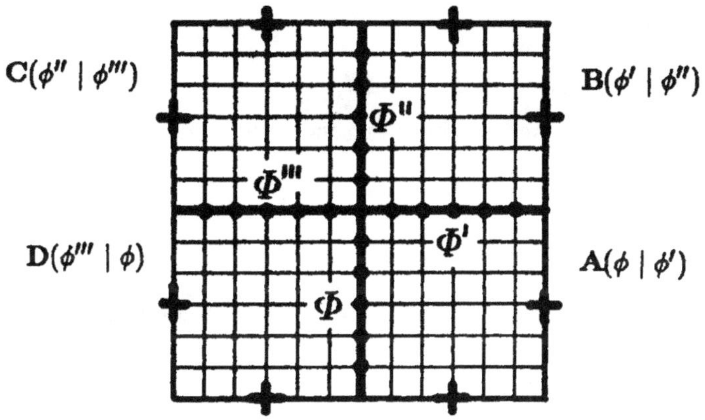

fig. 22 e

Then the partition function of the full lattice is (fig. 22 e):

$$Z_N = \sum_{\substack{all \\ spins}} \prod_{\substack{all \\ faces}} w(\sigma_i, \sigma_j, \sigma_k, \sigma_l) =$$

$$= \sum_\phi \sum_{\phi'} \sum_{\phi''} \sum_{\phi'''} A(\phi \mid \phi') B(\phi' \mid \phi'') C(\phi'' \mid \phi''') D(\phi''' \mid \phi) \qquad (3.44)$$

In a matrix notation,

$$\mathbf{A} = \{A(\phi \mid \phi')\}$$

and similarly for **B, C, D**, we have:

$$Z_N = \text{Tr} \, \mathbf{ABCD}$$

The summation in (3.44) is over all spin sets $\phi \cdots \phi'''$ subject to the restriction that $\sigma_1 = \sigma_1' = \sigma_1'' = \sigma_1'''$. This can be taken into account writing **A, B, C, D** in the following block diagonal form:

$$\mathbf{A} = \begin{array}{c} \\ + \\ - \end{array} \overset{\displaystyle + \qquad -}{\left(\begin{array}{c|c} /\!/\!/\!/ & 0 \\ \hline 0 & /\!/\!/\!/ \end{array} \right)} \qquad (3.45)$$

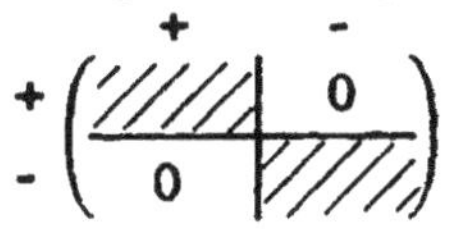

and the same for **B**, **C** and **D**.

Magnetization of eight-vertex model

We are interested in the mean value of the central spin, i.e.:

$$< \sigma_1 >= \frac{1}{Z_N} \sum_{\substack{all \\ spins}} \sigma_1 \prod_{\substack{all \\ faces}} w(\sigma_i, \sigma_j, \sigma_k, \sigma_l)$$

This can be rewritten using the corner transfer mtrices **A**, **B**, **C** and **D** in their block diagonal form (3.45):

$$< \sigma_1 >= \frac{\mathrm{Tr}\,\mathbf{SABCD}}{\mathrm{Tr}\,\mathbf{ABCD}} \tag{3.46}$$

$$\mathbf{S} = \begin{pmatrix} \mathbf{I} & 0 \\ 0 & \mathbf{I} \end{pmatrix}$$

Note that the position of **S** in the numerator product is irrilevant because it commutes with each of **A**, **B**, **C** and **D**.

The value of the magnetization $< \sigma_1 >$ (3.46) is unchanged by multiplying the corner matrices by scalar factor and also by similarity transformations:

$$\begin{aligned} \mathbf{A}' &= \mathbf{PAQ}^{-1} \\ \mathbf{B}' &= \mathbf{QBR}^{-1} \\ \mathbf{C}' &= \mathbf{RCS}^{-1} \\ \mathbf{D}' &= \mathbf{SDP}^{-1} \end{aligned}$$

That last property is very useful as it makes possible to use diagonal form for the corner matrices.

We can choose the scalar factor for **A**, **B**, **C** and **D** so that:

$$\mathbf{A} = (+ + \cdots + \mid + + \cdots +) = \mathbf{I}$$

and similarly for **B**, **C** and **D**; we also arrange the eigenvalues in decreasing order (the maximum being equal to one).

We work in the termodynamic limit ($m \to \infty$), using infinite dimensional matrices and eigenvectors and also infinite eigenvalues. In this limit the eigenvalues of the corner transfer matrix form a discrete set, while for example those of the row to row transfer matrix constitute a bond, i.e. they tend to a continuous set.

We compute, by means of the approach discussed before, the magnetization of the eight-vertex model.

Star triangle relations for the Boltzmann weights ensure that there are several symmetries for the corner transfer matrices. As **A**, **B**, **C**, **D** can be parametrized in terms of q, λ, u, we can set:

$$\mathbf{A} \equiv \mathbf{A}(u) \qquad \mathbf{B} \equiv \mathbf{B}(u) \qquad \mathbf{C} \equiv \mathbf{C}(u) \qquad \mathbf{D} \equiv \mathbf{D}(u)$$

Then:

$$\begin{aligned} \mathbf{C} = \mathbf{A} &= \mathbf{A}(u) \\ \mathbf{D} = \mathbf{B} &= \mathbf{A}(\lambda - u) \end{aligned} \tag{3.47}$$

Infact the bottom right and the top left corner transfer matrices are the same and so are the bottom left and the top right; besides we know that replacing u by $(\lambda - u)$ merely rotates everything of 90 degrees (it's the property (ii) of symmetry). We can use these symmetries of the weight function because the boundary conditions we gave (all spins up) don't break them; however, at least in the antiferromagnetic case and in the hard hexagon case these symmetries may be broken.

Using (3.47) the magnetization of the eight-vertex model (3.46) becomes:

$$M_0 = < \sigma_1 > = \frac{\mathrm{Tr}\, \mathbf{S}\mathbf{A}^2(u)\mathbf{A}^2(\lambda - u)}{\mathrm{Tr}\, \mathbf{A}^2(u)\mathbf{A}^2(\lambda - u)} \tag{3.48}$$

In this formula we've already used the fact that $\mathbf{A}(u)$ and $\mathbf{A}(\lambda - u)$ commute; it can be deduced in thermodynamic limit by means of (3.42). Infact because of it it's certainly true that:

$$\mathbf{A}(u)\mathbf{A}(v) = \text{scalar} \times \mathbf{X}(u + v) \tag{3.49a}$$

$$\mathbf{A}(u)\mathbf{A}(v) = \mathbf{A}(v)\mathbf{A}(u) \tag{3.49b}$$

Then, as we supposed in (3.48), $\mathbf{A}(u)$ and $\mathbf{A}(v)$ commute each other and with $\mathbf{X}(u + v)$: they can then be simultaneously diagonalized by a transformation which doesn't depend on u or v and also normalized so that their top left elements (in this case the maximum eigenvalue) becomes one:

$$\begin{aligned} \mathbf{A}_d(u) &= \frac{\mathbf{P}^{-1}\mathbf{A}\mathbf{P}}{\alpha_1(u)} \\ \mathbf{X}_d(u) &= \frac{\mathbf{P}^{-1}\mathbf{X}\mathbf{P}}{x_1(u)} \end{aligned}$$

We can then rewrite (3.49 a) in a diagonal form. The $i = 1$ case tells us that the scalar is 1; for $i \geq 2$ we have:

$$\alpha_i(u)\alpha_i(v) = x_i(u+v) \tag{3.50}$$

These equations must be true for all complex numbers u and v in some domain. They constitute a very strong condition on $\alpha_i(u)$ and infact it turns out that its general form has to be:

$$\alpha_i(u) = m_i e^{-r_i u} \tag{3.51}$$

where m_i, r_i are constants to be determined.

We have so gained with this approach much more informations about the eigenvalues of the corner tansfer matrix than in the case of the row to row transfer matrix, where from the commutation relations we obtained the eigenvalues as very complicated functions of u.

Now put in the magnetization (3.48) the diagonal form (3.51) of $\mathbf{A}$:

$$M_0 = \frac{\sum_i S_i m_i^2 e^{-2r_i u} m_i^2 e^{-2r_i(\lambda-u)}}{\sum_i m_i^2 e^{-2r_i u} m_i^2 e^{-2r_i(\lambda-u)}} = \frac{\sum_i S_i m_i^4 e^{-2r_i\lambda}}{\sum_i m_i^4 e^{-2r_i\lambda}}$$

$$S_i = \pm 1 \tag{3.52}$$

First one notes that u cancels out, which is the same result that was obtained using row to row transfer matrices.

We now determine m_i. Indeed it's generally true that when $u = 0$ $\mathbf{A}(u)$ is a diagonal matrix and for the ferromagnetic eight-vertex case $\mathbf{A}(0)$ is just the identity (multiplied by some scalar factor); so in this case $m_i = 1$.

Besides u enters the weight function only through snh u, i.e. through an infinite product of factor involving e^{2u}; so the weight function, being an analytic function of u, doesn't change incrementing u by $2\pi i$ and the same is true for the corner transfer matrix. That is a property of periodicity:

$$\mathbf{A}(u + 2\pi i) = \mathbf{A}(u)$$

We also need $\alpha_i(u)$ to be an analytic function of u in the strip between zero and λ in the complex plane of u. Because of this assumption r_i now must be integer. That's very interesting as we can calculate r_i in the low temperature limit and, r_i being continuous for $T \leq T_c$ and being an integer, the result must be true for all $T \leq T_c$.

So let's consider the low temperature limit. $\alpha_i(u)$ is certainly a product over all faces of the bottom right corner lattice of face's weight functions. We immagine to construct this product beginning with the top left corner face; since the interaction has a pratically diagonal form $\sigma_i\sigma'_i \sim \delta(\sigma_i, \sigma'_i)$ in the low temperature limit, the corner lattice can be drawn as in fig. 23.

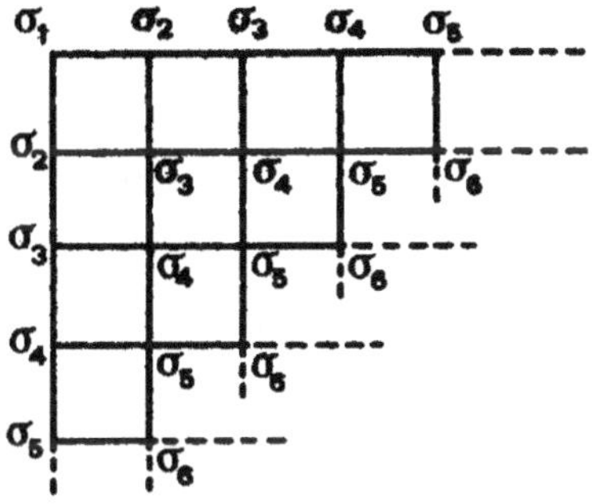

fig. 23

Then:

$$\prod_{\substack{all\\faces}} w(\sigma_i, \sigma_{i+1}, \sigma'_i, \sigma_{i-1}) = ce^{\lambda(\sigma_1\sigma_3 + 2\sigma_2\sigma_4 + \cdots + m\sigma_m\sigma_{m+2})}$$

It is rather obvious now that $\alpha_i(u)$ in the low temperature limit turns out to be:

$$\alpha_i(u) = \exp\{-u(\sigma_1\sigma_3 + 2\sigma_2\sigma_4 + \cdots + m\sigma_m\sigma_{m+2})\}$$

$$\sigma_{m+1} = \sigma_{m+2} = +1$$

Thus:

$$r_i = \sigma_1\sigma_3 + 2\sigma_2\sigma_4 + \cdots + m\sigma_m\sigma_{m+2} \tag{3.54}$$

Finally we set:

$$S_i = \sigma_i$$

Spontaneous magnetization then becomes:

$$M_0 = \frac{\sum_{\sigma_1\cdots\sigma_m} \sigma_1 e^{-2\lambda(\sigma_1\sigma_3 + 2\sigma_2\sigma_4 + \cdots + m\sigma_m\sigma_{m+2})}}{\sum_{\sigma_1\cdots\sigma_m} e^{-2\lambda(\sigma_1\sigma_3 + 2\sigma_2\sigma_4 + \cdots + m\sigma_m\sigma_{m+2})}} \tag{3.55}$$

The expression (3.55) suggests that the partition function is really like that of a one dimensional model but with interaction coefficients increasing with position. To compute the sum in (3.55) one has only to go from the set of spins $\{\sigma\}$ to a new set of spins τ defined by:

$$\tau_j = \sigma_j \sigma_{j+2} \qquad\qquad j = 1, \ldots, m$$
$$\sigma_1 = \tau_1 \tau_3 \cdots \tau_m$$

Then:

$$M_0 = \frac{\sum_{\tau_1 \cdots \tau_m} \tau_1 \tau_3 \cdots \tau_m e^{-2\lambda(\tau_1 + 2\tau_2 + \cdots + m\tau_m)}}{\sum_{\tau_1 \cdots \tau_m} e^{-2\lambda(\tau_1 + 2\tau_2 + \cdots + m\tau_m)}} \tag{3.56}$$

This sum now factors into a sum over τ_1 times a sum over τ_2 ... and so on. The even terms in the numerator and in the denominator cancel out and one is left with the odd terms; setting $x = e^{-2\lambda}$ we find:

$$M_0 = \frac{1-x^2}{1+x^2} \frac{1-x^6}{1+x^6} \frac{1-x^{10}}{1+x^{10}} \cdots \tag{3.57}$$

In the limit $m \to \infty$ this expression converges very nicely. This infinite product is just an elliptic function of x.

The result (3.57) was conjectured in 1973 by M. Barber and R. Baxter [20]. The basis argument was that M_0 had to be indipendent on u and also on q; so taking the known result for M_0 in the Ising model, which corresponds to $q = x^2$ they were able to write down the general eight-vertex model. The result was then proved in 1980 [21].

Local density for hard hexagon model

We consider again the average of the central spin σ_1, remembering that in this case:

$$\sigma_i = 0, 1 \qquad\qquad \sigma_i \sigma_{i+1} = 0 \qquad\qquad 1 \le i \le m \tag{3.58}$$

besides we fix: $\sigma_{m+1} = 0$.

We'll again use (3.51) with (3.52) but now $\mathbf{A}(u)$ for $u = 0$ is no more proportional to the identity; it is however still diagonal, depending only on σ_1.

Following arguments similar to those used for eight-vertex model case, we finally find for $\alpha_i(u)$:

$$\alpha_i(u) = R^{\frac{1}{2}} \sigma_i e^{-3u(\sigma_2 + 2\sigma_3 + \cdots + (m-1)\sigma_m)} \tag{3.59}$$

To explain the meaning of R let's introduce two elliptic functions:

$$G(x) = \left[(1-x)(1-x^4)(1-x^6)(1-x^9)(1-x^{11})(1-x^{14})\cdots\right]^{-1} \quad (3.60a)$$

$$H(x) = \left[(1-x^2)(1-x^3)(1-x^7)(1-x^8)(1-x^{12})(1-x^{13})\cdots\right]^{-1} \quad (3.60b)$$

$$x = e^{-\lambda}$$

Then R is actually given by:

$$R^2 = \frac{-xG(x)}{H(x)} \qquad (-1 < x < 0) \qquad (3.61)$$

Now substitute (3.58) into the general formula (3.52) for the local density, where as for the eight-vertex model $S_i = \sigma_i$:

$$\rho = <\sigma_1> = \frac{\sum_{\sigma_1 \cdots \sigma_m} \sigma_1 R^{2\sigma_1} e^{-6\lambda(\sigma_2 + 2\sigma_3 + \cdots + (m-1)\sigma_m)}}{\sum_{\sigma_1 \cdots \sigma_m} R^{2\sigma_1} e^{-6\lambda(\sigma_2 + 2\sigma_3 + \cdots + (m-1)\sigma_m)}} \qquad (3.62)$$

We can compute the sum in σ_1 obtaining:

$$\rho = \frac{R^2 F(1)}{F(0) + R^2 F(1)} \qquad (3.63)$$

$$F(\sigma_1) = \sum_{\sigma_2 \cdots \sigma_m} q^{\sigma_2 + 2\sigma_3 + \cdots + (m-1)\sigma_m} \qquad (3.64)$$

$$q = e^{-6\lambda}$$

We now calculate, by means of a sort of perturbation approach, the summation in (3.64), subject to the rule (3.58). Take $\sigma_2 = \cdots = \sigma_m = 0$ as the ground state: it gives a contribution one both to $F(0)$ and to $F(1)$. Then allow one of the $\{\sigma_r\}$ to be one and sum over all the possibilities, remembering that while for $F(0)$ σ_2 may be 1, for $F(1)$ the value 1 isn't allowed for the rule (3.58). Then allow two of the $\{\sigma_r\}$ to be one, and so on. With a straightforward calculation one obtains:

$$F(0) = 1 + \frac{q}{1-q} + \frac{q^4}{(1-q)(1-q^2)} + \frac{q^9}{(1-q)(1-q^2)(1-q^3)} + \cdots$$

$$F(1) = 1 + \frac{q^2}{1-q} + \frac{q^6}{(1-q)(1-q^2)} + \frac{q^{12}}{(1-q)(1-q^2)(1-q^3)} + \cdots$$

In about 1890 Rogers showed that these series are just the elliptic functions defined in (3.60) (Rogers-Ramanujan identities)[22, 23]:

$$\begin{aligned} F(0) &= G(q) \\ F(1) &= H(q) \end{aligned} \tag{3.65}$$

Substitute these results in (3.63), using (3.61):

$$\rho = \frac{-xG(x)H(x^6)}{H(x)\left\{G(x^6) - x\frac{G(x)}{H(x)}H(x^6)\right\}} \tag{3.66}$$

From another of the Rogers-Ramanujan identities we have:

$$H(x)G(x^6) - xG(x)H(x^6) = \frac{P(x)}{P(x^3)}$$

where:

$$P(x) = \prod_{n=1}^{\infty} \left(1 - x^{2n-1}\right)$$

So the denominator in (3.66) simplifies and finally we get the local density in terms of elliptic function products:

$$\rho = \frac{-xG(x)H(x^6)P(x^3)}{P(x)} \tag{3.67}$$

This result holds only for densities $\rho \leq \rho_c$, when q lies between 0 and 1. There are many other phases of the hard hexagon model, in which however one uses identities like these (there are about 40 identities listed by Ramanujan).

DUALITY

It's possible to convert the Ising spin model on a square lattice to a vertex model if the Boltzmann weight is an even function of the four spin round a face:

$$w\begin{pmatrix} d & c \\ a & b \end{pmatrix} = w\begin{pmatrix} -d & -c \\ -a & -b \end{pmatrix}$$

In that case instead of specifying w by the values of a, b, c, d, one could specify it by the values of their products along the four edges (fig. 24):

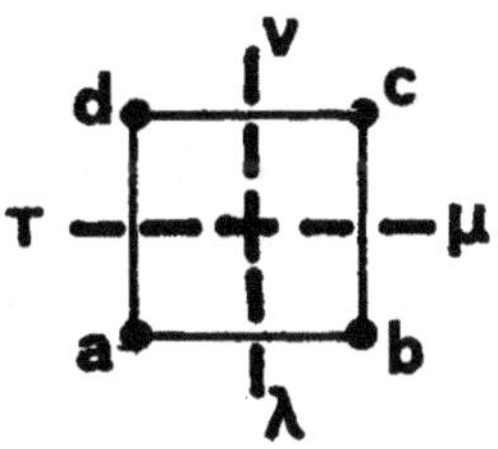

$\lambda = ab$ $\qquad \mu = bc$ $\qquad \nu = cd$ $\qquad \tau = da$

fig. 24

Then:

$$w = w(\lambda, \mu, \nu, \tau), \qquad \lambda\mu\nu\tau = 1 \tag{4.1}$$

Because of this last restriction there are only eight possible values of λ, μ, ν and τ; this is the general eight-vertex model in a field, which is an unsolved problem.

In fig. 24 the dot lines form the dual lattice of the continuous lines. If we rotate this dual lattice of 45 degrees we obtain fig. 25 which can be thought of as a sort of process in which two incoming arrows (λ and μ) go into two outgoing ones (τ and ν).

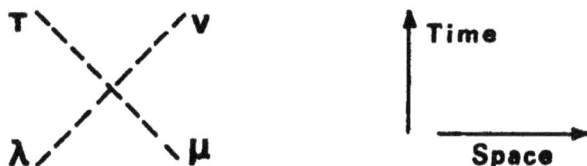

fig. 25

If we take the axis like in fig. 25 b, that is a scattering process in $1+1$ dimensions and w from that point of view is a scattering matrix, say an *S matrix*. Set:

$$w\begin{pmatrix} d & c \\ a & b \end{pmatrix} = S\begin{pmatrix} \tau & \nu \\ \lambda & \mu \end{pmatrix} \tag{4.2}$$

Star triangle relations (3.29) become:

$$\begin{aligned} &\sum_{\lambda,\mu,\nu} S\begin{pmatrix} \alpha & \lambda \\ \beta & \mu \end{pmatrix} S'\begin{pmatrix} \epsilon & \nu \\ \phi & \lambda \end{pmatrix} S''\begin{pmatrix} \gamma & \mu \\ \delta & \nu \end{pmatrix} = \\ &= \sum_{\lambda,\mu,\nu} S''\begin{pmatrix} \nu & \alpha \\ \mu & \phi \end{pmatrix} S'\begin{pmatrix} \lambda & \gamma \\ \nu & \beta \end{pmatrix} S\begin{pmatrix} \mu & \epsilon \\ \lambda & \delta \end{pmatrix} \end{aligned} \tag{4.3}$$

The sum is over all values of λ, μ, ν subject to the condition that the product of the edge spins round each vertex is one. There will only be two such possibilities and one of them being known one can always obtain the other by negating λ, μ and ν. So the sum in (4.3) has only two terms, as originally (see (3.29)) for the sum over the central spin g.

Again we can draw (4.3) pictorially:

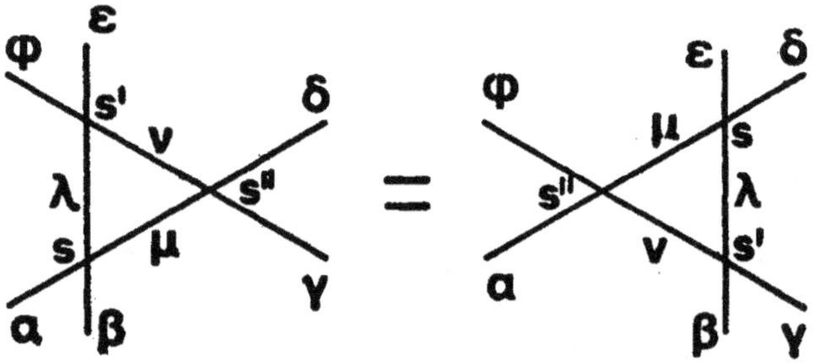

Star triangle relation (4.3) can be recognized in field theory as the factorizability condition of the S matrix. That is the way, due to Zamolodchikov [24], to make a contact with the field theoretic point of view.

THREE DIMENSIONS

In three dimensions one defines a model analogous to the IRF one in two dimensions; we now have an ***interaction round a cube*** model, on a cubic lattice. In an obvious way we generalize the partition function:

$$Z = \sum_{\{\sigma\}} \prod_{\substack{cubes \\ (i,j,k,l \\ p,q,r,s)}} w(\sigma_i \mid \sigma_p \sigma_q \sigma_r \mid \sigma_j \sigma_k \sigma_l \mid \sigma_s) \tag{5.1}$$

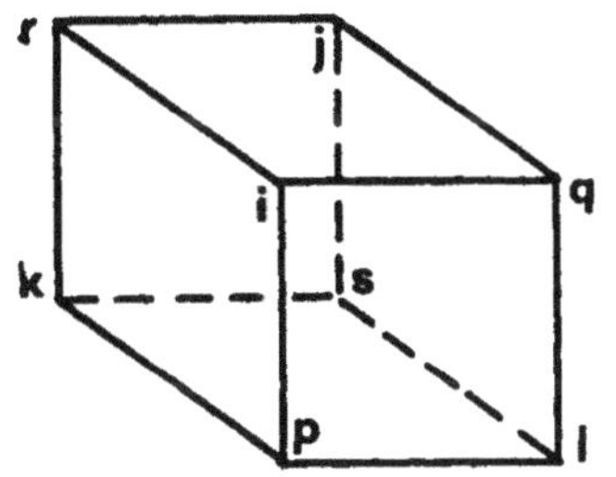

fig. 26

Here w is the Boltzmann weight function of the cube (fig. 26) and depends on the eight spins round the cube. As each spin has two possible values there are 2^8 possible configurations of spins on the cube, so w depends on 256 parameters.

To simplify the notation set:

$$w = w(a \mid bcd \mid efg \mid h)$$

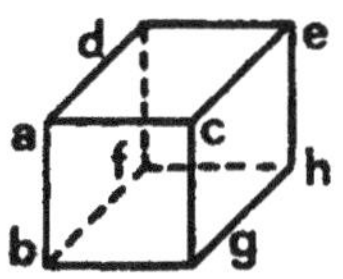

We can now generalize star triangle relations in three dimensions: it is the condition for two layer to layer transfer matrices to commute [25]. Star triangle relations become the ***Tetrahedron*** relations (note that both of these terms really apply to the dual system):

$$\sum_g w(d \mid vuw \mid prq \mid q)w'(u \mid qcp \mid xgz \mid s) \times$$
$$\times w''(p \mid gxw \mid brs \mid y)w'''(g \mid qsr \mid yvz \mid a) =$$
$$= \sum_g w'''(p \mid uxv \mid bcd \mid g)w''(u \mid qcd \mid gvz \mid a) \times \tag{5.2}$$
$$\times w'(d \mid vgw \mid bra \mid y)w(g \mid acb \mid xyz \mid s)$$

This relation has to hold for all values of the fourteen external spins $a\,b\,c\,d$, $p\,q\,r\,s$, $u\,v\,w\,x\,y\,z$, so there are 2^{14} equations. On the other hand there are only four weight functions, each depending on 2^8 parameters, so only $4 \times 2^8 = 2^{10}$ equations are unknown.

Let's give a graphical interpretation of the tetrahedron relations. We consider for example the left hand side of (5.2) as the partition function of a graph and the graph looks just like in fig. 27.

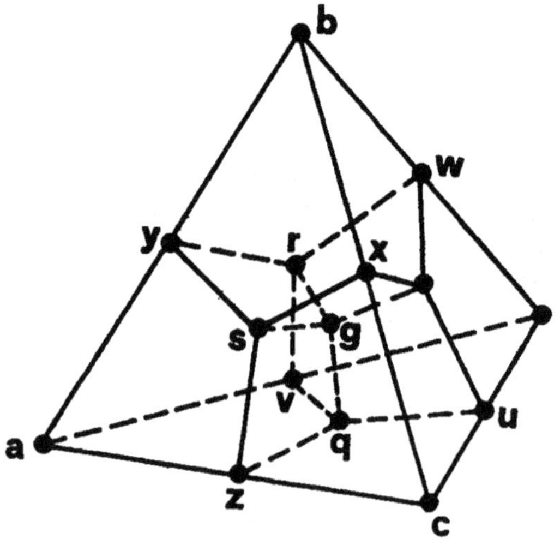

fig. 27

Note that considering say the spins $a\,z\,q\,g\,s\,y\,r\,v$ connected with each other, they form a (distorted) cube whose weight function is w''.

For the right hand side of (5.2) a similar graphic rapresentation is true, in which g instead of being connected to the centers of the faces (fig. 27) is connected to the vertices.

We want to solve the 2^{14} equations (5.2), 2^{10} of them being unknown. First we can think to look for a solution that factors into two distinct sublattices, in analogy to the Ising model case of the eight-vertex model in two dimensions.

Unfortunately odd dimensions are harder than even. While in two dimensions each equation factors into two identical equations, in three dimensions it factors into two equations that are different, one of them being trivial and the other having a sum on both sides.

Nevertheless Zamolodchikov, using field theoretic intuitions and the dual S matrix picture, conjectured a solution to these 2^{14} equations [26].

He tested it in various ways and did a numerical test of the equations on computer. Baxter then gave a general proof [27].

The model is specified by table 5:

$abeh$	$acfh$	$adgh$	$W(a \mid efg \mid bcd \mid h)$
+	+	+	$P_0 - abcdQ_0$
−	+	+	R_1
+	−	+	R_2
+	+	−	R_3
+	−	−	$abP_1 + cdQ_1$
−	+	−	$acP_2 + bdQ_2$
−	−	+	$adP_3 + bcQ_3$
−	−	−	R_0

table 5

In this model to fix w is sufficient to specify the values of the products $a\,b\,e\,h$, $a\,c\,f\,h$, $a\,d\,g\,h$, where as usual we refer to the cube:

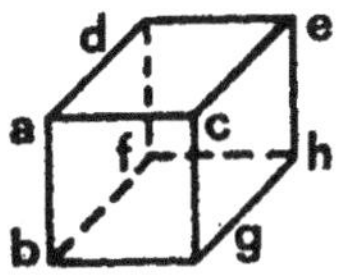

There are 256 configurations, 16 of them having weight function $w(+,+,+)$, 16 of them having weight function $w(-,+,+)$ and so on.

In this model there are several symmetries:

(i) The weight function is unchanged by reversing all spins:

$$w(-a \mid -e-f-g \mid -b-c-d \mid -h) = \\ w(a \mid efg \mid bcd \mid h)$$

(ii) The weight function is unchanged by reversing half of the spins (forming a tetrahedron):

$$w(-a \mid efg \mid -b-c-d \mid h) = \\ = w(a \mid -e-f-g \mid bcd \mid -h) = \\ = w(a \mid efg \mid bcd \mid h)$$

(iii) If one reverses all the spins on one face the weight function either remains unchanged or changes sign:

$$w(-a \mid -e-f-g \mid bc-d \mid h) = \\ = \pm w(a \mid efg \mid bcd \mid h)$$

Here the antisymmetry relations mean that the Zamolodchikov model can have negative Boltzmann weights so it is unphysical from the point of view of statistical mechanics (perhaps it doesn't matter from the point of view of field theory). Even so, as we'll see, it is a solution of the tetrahedron relations and any non trivial solution is interesting. Infact as in two dimensions corner transfer matrices provide a very powerful method of developing approssimate solutions of unsolved models, so if we can understand this model and what for this model are the analogues of two dimensional corner transfer matrices then probably we'll gain some information about unsolved three dimensional models

(iv) Finally if one interchanges a with p, b with q, c with r, d with s the left hand and right hand side of tetrahedron relations are interchanged too.

So 2^{10} equations reduce to 224. We now need to specify the twelve constants $P_0, \ldots, P_3, R_0, \ldots, R_3, Q_0, \ldots, Q_3$ (real numbers). Consider the spherical triangle in fig. 28:

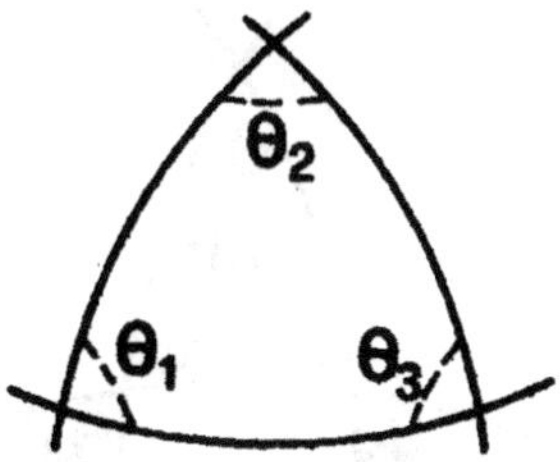

fig. 28

and define:

$$\alpha_0 = \frac{1}{2}(\theta_1 + \theta_2 + \theta_3 - \pi)$$
$$\alpha_i = \frac{1}{2}(\pi + \theta_i - \theta_j - \theta_k)$$

for all permutation i, j, k of $(1, 2, 3)$. Also:

$$t_i = \sqrt{\tan\frac{\alpha_i}{2}} \qquad c_i = \sqrt{\cos\frac{\alpha_i}{2}} \qquad s_i = \sqrt{\sin\frac{\alpha_i}{2}}$$

Then Zamolodchikov solution is:

$$\begin{array}{lll} P_0 = 1 & Q_0 = t_0 t_1 t_2 t_3 & R_0 = \frac{s_0}{c_1 c_2 c_3} \\ P_i = t_j t_k & Q_j = t_0 t_i & R_i = \frac{s_i}{c_0 c_j c_k} \end{array} \tag{5.3}$$

Regarding these equations as defining w as a function of $\theta_1, \theta_2, \theta_3$, Zamolodchikov solution is that if w has angles $\theta_1, \theta_2, \theta_3$, then w', w'', w''' are defined similarly but with angles:

$$\begin{array}{ll} w': & \pi - \theta_6, \theta_2, \pi - \theta_4 \\ w'': & \theta_5, \pi - \theta_3, \pi - \theta_4 \\ w''': & \theta_5, \theta_1, \theta_6 \end{array} \tag{5.4}$$

where $\theta_1, \ldots, \theta_6$ are the six angles of the spherical quadrilateral shown in fig. 29; five of them are independent while the 6th is determined by the rest.

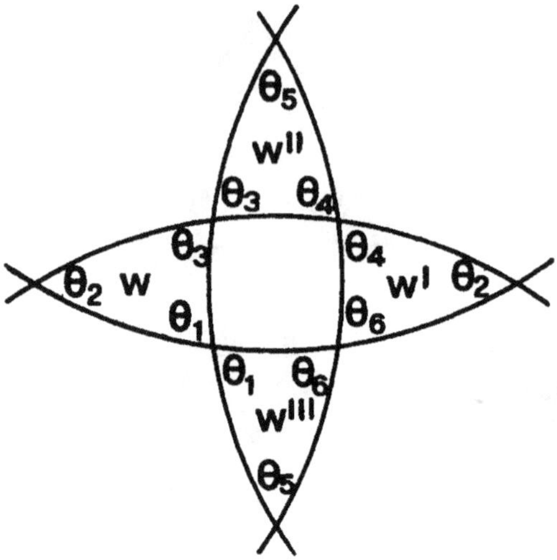

fig. 29

If Zamolodchikov solution is true each of the 224 tetrahedron relations should be an identity for all the allowed values of the angles. It turns out that different equations can be obtained from one another by permuting $\theta_1, \ldots, \theta_6$, or rearranging the spherical quadrilateral.

Using this, the 224 equations reduce to only two identities and we have to prove just these two: they are true.

Next step would be to determine the partition function per site $k(\theta_1, \theta_2, \theta_3)$, that has still not been obtained. Infact we do have an inversion relation but this determines k only if appropriate analyticity assumptions are made and it is not obvious what these are. Besides w doesn't include a temperature-like variable, so we are not able to check the analyticity assumption with high-temperature or low-temperature expansions (as for two dimensional solved models).

Our guess is that the model is always critical and varying $\theta_1, \theta_2, \theta_3$, merely moves w around the critical surface: $\theta_1, \theta_2, \theta_3$, are irrelevant variables. The reasons for this guess are exentially the following:

(i) If we look at the two-layer Zamolodchikov model it turns out that it is equivalent to the two dimensional free fermion model [28], which is critical

(ii) Baxter and Forester have generalized the Zamolodchikov model to allow a ***temperature*** variable T. From a numerical approximation they have found that the temperature corresponding to Zamolodchikov model is at least very near to T_c (fig. 30).

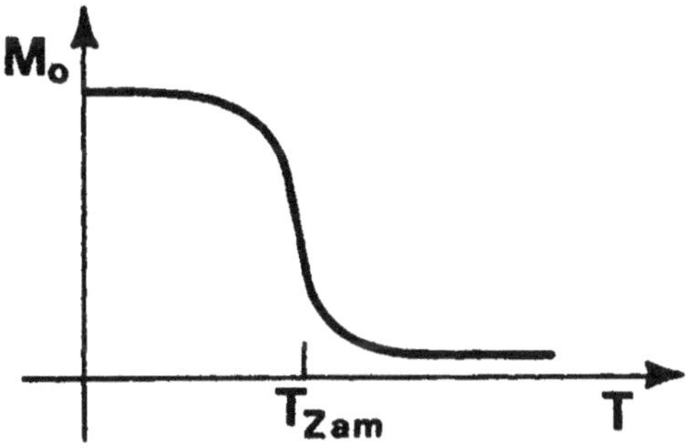

fig. 30

(iii) in 2-D, the Potts model can be solved only at criticality.

We conclude these lectures with the convintion that Zamolodchikov model can actually be written as a three spin model on the triangles obtained by adding a central spin in the cubes of the cubic lattice and connecting it to all eight vertex of the cube as in fig. 31:

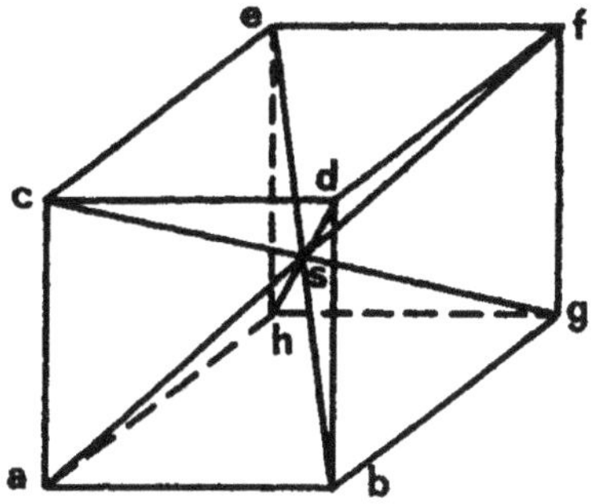

fig. 31

We have to use only six of the twelve triangles. Then the weight function of the cube is a sum over the central spin S of the product of the weight functions of the six triangles, each of them being given by (say for the triangle $c\,d\,e$):

$$w = e^{\kappa \sigma_c \sigma_d \sigma_e} \tag{5.5}$$

where κ is the appropriate coefficient.

References

[1] L. Tonks, Phys. Rev. **50**, 955 (1936).
[2] E. Ising, Z. Physik **31**, 253-8 (1925).
[3] H. Takahashi, Proc. Phys.-Math. Soc. Japan (Nippon Suugaku-Buturigak-Kwai Kizi, Tokyo)**24**, 60 (1942).
[4] L. van Hove, Physica **16**, 137-43 (1950).
[5] W.L. Bragg, E.J. Williams, Proc. Roy. Soc. (London)**A145**, 699-730 (1934).
[6] H.A. Bethe, Proc. Roy. Soc. (London)**A150**, 552-75 (1935).
[7] M. Kac, G.E. Uhlenbeck and P.L. Hemmer, J. Math. Phys. **4**, 216-247 (1963); J. Math. Phys. **5**, 60-84 (1964).
[8] T.H. Berlin and M. Kac, Phys. Rev **86**, 821-35 (1952).
[9] R.J. Baxter, J. Phys. A: Math. Gen. **13**, L61-L70 (1980).
[10] L. Onsager, Phys. Rev. **65**, 117-49 (1944).
[11] P.W. Kasteleyn, Physica **27**, 1209-25 (1961).
[12] M.E. Fisher, Phys. Rev. **124**, 1664-72 (1961).
[13] E.H. Lieb, Phys. Rev. **162**, 162-72 (1967).
[14] E.H. Lieb, Phys. Rev. Lett. **18**, 1046-8 (1967).
[15] E.H. Lieb, Phys. Rev. Lett. **19**, 108-10 (1967).
[16] R.J. Baxter, Phys. Rev. Lett. **26**, 832-3 (1971).
[17] R.J. Baxter, Phys. Rev. Lett. **26**, 834 (1971).
[18] R.J. Baxter and I.G. Enting, J. Phys. A: Math. Gen **9**, L149-52 (1976).
[20] F.G. Frobenius, Reprinted in Sect. 79 of "Ferdinand G. Frobenius: Gesammelte Abhandlungen", Springer-Verlag, Berlin, 1968
[21] M.N. Barber and R.J. Baxter, J. Phys. C: Solid State Phys. **6**, 2913-21 (1973).
[22] R.J. Baxter, Exactly Solved Models in Statistical Mechanics, Academic Press London, (1982).
[23] L.J. Rogers, Proc. Lond. Math. Soc. **25**, 318-43 (1894).
[24] S. Ramanujan, Proc. Camb. Phil. Soc. **19**, 214-6 (1919).
[25] A.B. Zamolodchikov, Letter to Zh. Eksper. Teoret. Fiz. **25**, 499-502 (1977).
[26] Jackel and Maillard, J. Phys. A: Math. Gen. **15**, 1309 (1982).
[27] A.B. Zamolodchikov, Commun. Math. Phys. **79**, 489-505 (1981).
[28] R.J. Baxter, Comm. Math. Phys. **88**, 2,185 (1983).
[29] C. Fan and F.Y: Wu, Phys. Rev **B2**, 723-33 (1970).

Integrable Systems and Infinite Dimensional Lie Algebras

MICHIO JIMBO AND TETSUJI MIWA

Research Institute for Mathematical Sciences
Kyoto University
Kyoto 606, Japan

Contents

1. INTRODUCTION

By "Integrable Systems" we mean two different branches: one is that of classical systems, i.e. "Non linear partial differential equations" which are called by physicists "soliton equations"; the other branch is that of quantum systems, in statistical mechanics and field theory: the typical example is the Ising model. Specifically we want to emphasize the role of infinite dimensional Lie algebras, or translated in other words, the transformation groups of the system (the hidden symmetry group). In the first part of the paper we will be concerned mainly with the Ising model and the related monodromy problem [1]; in the second part we will treat solitons and the trasformation groups theory. To give an ideà of the type of results which we are going to treat, we will show the most important one, which is related to the 2-dim. Ising model. It is well known that the Ising model is one of the very few solvable models for which all the correlation functions are known.

Denoting by $<\sigma_{00}\sigma_{MN}>$ the two-point correlation function on the lattice, let us take the scaling limit $T \to T_c \pm 0$ (T_c being the critical temperature), mantaining the products $\epsilon M = x_1$, $\epsilon N = x_2$ fixed ($\epsilon = |T - T_c|$) then by a proper scaling the correlation function will tend to a function depending on the continuous variables x_1, x_2:

$$\lim_{T \to T_c \pm 0} \epsilon^{-1/4} <\sigma_{00}\sigma_{MN}> = \tau_{\pm}(x) \qquad x = (x_1, x_2) \tag{1.1}$$

Then put:

$$u(x) = -\frac{1}{2} \ln \frac{\tau_{-}(x) - \tau_{+}(x)}{\tau_{-}(x) + \tau_{+}(x)} \tag{1.2}$$

The result is that $u(x)$ satisfies the Sinh-Gordon equation (one of the most typical examples of soliton equations):

$$\Delta u = \sinh 2u \tag{1.3}$$

Of course $u(x)$ is only a particular solution of the Sinh-Gordon equation. It turns out that $\tau_{\pm}(x)$ and $u(x)$ are rotationally symmetric, i.e. functions only of $t = |x| = \sqrt{x_1^2 + x_2^2}$. From (1.2) we obtain a non linear ordinary differential equation (NLODE):

$$\left(\frac{d^2}{dt^2} + \frac{1}{t}\frac{d}{dt}\right)u = \sinh 2u \tag{1.4}$$

which is equivalent to a Painlevé equation of the 3rd kind. This is the remarkable result of Wu-McCoy-Tracy-Barouch [2, 3, 4]. This example shows all the subjects which we are concerned with: models in statistical mechanics, soliton equations and the third one, which is the monodromy problem that we are going to explain.

2. MONODROMY AND RIEMANN'S PROBLEM

Painlevé is a French mathematician who worked at the beginning of the century. His most significant work was devoted to classification of the 2nd order NLODE which have the property that the general solution has no movable branch points. In the case of linear differential equations (LDE), the singularity of the solutions is automatically determined by looking at the equation: it derives from the singularity in the coefficients. On the contrary, in NLDE the singularity may depend on the integration constants as in the following example:

$$2yy' = 1 \qquad y = \sqrt{x - C} \tag{2.1}$$

The example shows a movable branch point, i.e. a branch point which depends on the integration constants. On the other hand, the general solution to the 1st order equation

$$(y')^2 = y^3 - g_2 y - g_3 \tag{2.2}$$

is given by $y = \wp\,(x - C)$ (elliptic function), whose movable singularities are only poles.

Painlevé tried to classify also the 2nd order NLDE having no movable branch point and obtained six new canonical types (which we refer to as PI-PVI) apart from the ones that can be solved in term of elliptic functions or solutions of LDE. For example, the simplest looks:

$$y'' = 6y^2 + x \tag{2.3}$$

which can be compared to (2.2) once differentiated:

$$y'' = 6y^2 - \frac{1}{2}g_2 \tag{2.4}$$

Painlevé equations have also another origin which at first sight is totally different from Painlevé's viewpoint: it is the Riemann's monodromy problem. This is related to linear ordinary differential equations (LODE). Let us consider a system of m LODE:

$$\frac{d}{dx}\mathbf{y} = A(x)\mathbf{y} \qquad \mathbf{y} = \begin{pmatrix} y_1 \\ \vdots \\ y_m \end{pmatrix} \tag{2.5}$$

$A(x)$ is an $m \times m$ matrix with rational coefficients, $x \in \mathbb{C}$; e.g., in the simplest case, if A has only simple poles:

$$A(x) = \sum_{\nu} \frac{A_\nu}{x - a_\nu} \qquad A_\nu = \text{ constant matrices} \tag{2.6}$$

Now let us suppose to have m linearly independent solutions $\mathbf{y}^{(j)}$ of the equation. We take a point x_0 as a reference point and consider the analytic continuation of these independent solutions along some path γ: starting from x_0 we move around, avoiding the singular points, coming back to the original point like in fig.2.1:

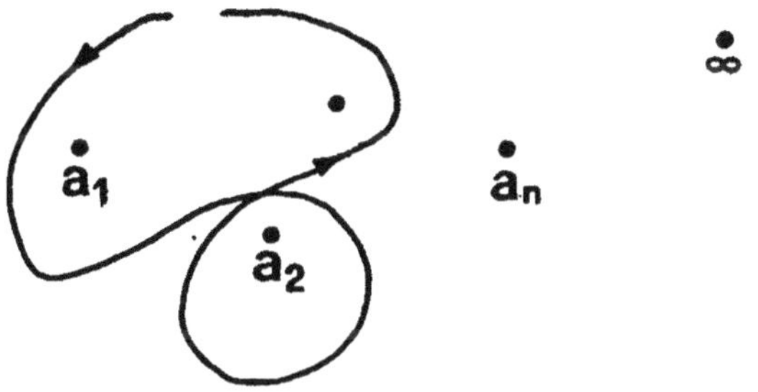

fig. 2.1

Since the coefficients of the differential equations are single valued functions, the analytical continued solutions should be again solutions of the equation, but they can be different, so the result is a linear transformation:

$$\mathbf{y}^{(j)} \xrightarrow{\gamma} \sum_{i=1}^{m} \mathbf{y}^{(i)} M_{\gamma,ij} \tag{2.7}$$

or in matrix notation:

$$Y(x) \xrightarrow{\gamma} Y(x) M_\gamma \tag{2.8}$$

$$Y(x) \equiv \left(\mathbf{y}^{(1)}, \ldots, \mathbf{y}^{(m)}\right), \qquad M_\gamma \equiv (M_{\gamma,ij})$$

M_γ is a constant matrix: it must depend only on the homotopy class of the path, i.e. if we change the path slightly, without meeting branch points, then the result should be the same. Another property of the matrices M_γ is that they must satisfy the composition law of the corrisponding paths: $M_\gamma M_{\gamma'} = M_{\gamma\gamma'}$ (i.e. if one goes around the path γ, then follows another path γ', the analytical continuation should be the same of the resulting path $\gamma\gamma'$). This means that the collection of matrices $\{M_\gamma\}$ form a group; this is the "monodromy group" of the equation. In other words, one has the representation of the homotopy group

$$\begin{array}{ccc}\pi_1(\mathbf{P}^1 - \{a_1, \ldots, a_n, \infty\}, x_0) & \longrightarrow & \mathcal{GL}(m, \mathbb{C}) \\ [\gamma] & \longrightarrow & M_\gamma\end{array}$$

Briefly we say that the monodromy group is a measure of the multivalued nature of the function $Y(x)$. So from a LDE follows a monodromy group. Riemann's problem is to ask whether the converse is true: given a matrix group $\mathcal{M} = \{M_\gamma\}$ find the $Y(x)$ that has $\mathcal{M}$ as its monodromy group. It is stated more precisely as follows. Having the LDE of the form

$$\frac{dY}{dx} = \sum_{\nu=1}^{n} \frac{A_\nu}{x - a_\nu} Y \tag{2.9}$$

one can show that its fundamental solution matrix Y has the properties:

(i)Y is holomorphic invertible for $x \neq a_\nu$

(ii)at $x = a_\nu$ it has the form:

$$Y(x) = \hat{Y}^{(\nu)}(x)(x - a_\nu)^{L_\nu} \tag{2.10}$$

$\hat{Y}^{(\nu)}(x)$ holomorphic invertible at $x = a_\nu$

for some L_ν ($\nu = 1, \ldots, n$). The monodromy matrices themselves are related to the exponents L_ν as

$$M_{\gamma_\nu} = e^{2\pi i L_\nu} \tag{2.11}$$

where the path γ_ν starting and ending at x_0 surrounds a_ν one time in the anticlockwise direction. M_{γ_ν} are determined uniquely by L_ν (but the converse is not true: for instance one can shift all the L_ν by integers, the group $\mathcal{M} = \{M_\gamma\}$ remaining unchanged). Riemann's problem is to find a $Y(x)$ satisfying (2.10) for given $L_1, \ldots, L_n$. One imposes also the normalization $Y(x_0) = 1$. From this definition of Riemann's problem it is very easy to see that if the solution exists it must be unique; on the other hand it is known that at least if all the L_ν are small the solution does exist. Riemann showed that the solution necessarily satisfies a LDE of the form (2.9). The argument is very simple: the matrix product $dY/dx \cdot Y^{-1}$ turns out to be single valued and has only simple poles in a_ν as $\hat{Y}^{(\mu)}(x)$ is holomorphic and invertible at $x = a_\mu$ and the analytic continuation given by equation (2.8) does not change $dY/dx \cdot Y^{-1}$ as the matrices M_γ and M_γ^{-1} cancel in the product.

Riemann asked also another question: suppose the monodromy group is fixed (i.e. the exponents L_ν are fixed) and consider A_ν as a function of L_ν and a_ν:

$$\begin{aligned} A_\nu &= A_\nu(a_1,\ldots,a_n;L_1,\ldots,L_n) \\ Y(x) &= Y\left(x;\begin{matrix} a_1,\ldots,a_n \\ L_1,\ldots,L_n \end{matrix}\right) \end{aligned} \tag{2.12}$$

What happens by varying $a_1,\ldots,a_n$? This is a problem of monodromy preserving deformation, i.e. the LODE (2.9) is deformed by varying $a_1,\ldots,a_n$ while the monodromy group is unchanged: Schlesinger [5] gave the following answer to this question. Taking $x_0 = \infty$, Y satisfies two LDE:

$$\begin{aligned} \frac{d}{dx}Y &= \sum_\nu \frac{A_\nu}{x-a_\nu}Y \\ \frac{\partial}{\partial a_\nu}Y &= -\frac{A_\nu}{x-a_\nu}Y \end{aligned} \tag{2.13}$$

and hence the compatibility condition:

$$\frac{\partial}{\partial a_\nu}\frac{\partial}{\partial x}Y = \frac{\partial}{\partial x}\frac{\partial}{\partial a_\nu}Y \tag{2.14}$$

leads to a non linear system of partial differential equations known as "Schlesinger equations":

$$\frac{\partial A_\mu}{\partial a_\nu} = \begin{cases} \frac{-[A_\mu,A_\nu]}{a_\mu-a_\nu} & (\mu \neq \nu) \\ \sum_{\lambda(\neq\nu)} \frac{[A_\lambda,A_\nu]}{a_\lambda-a_\nu} & (\mu = \nu) \end{cases} \tag{2.15}$$

The equations (2.15) are so the necessary and sufficient conditions for the monodromy to be preserved, i.e. they are the monodromy preserving deformation equations. The equations (2.15) are connected to the Painlevé equations. In the simplest case the size of the matrices A_γ is 2×2. If the number of poles is three (including infinite) the hypergeometric equation is obtained; if there are four poles (which can be put $a_1 = 0$, $a_2 = 1$, $a_3 = t$, $a_0 = \infty$) after some elimination one can get a NLODE of the second order with respect to the variable t which is the Painlevé equation of the VI kind. All the other Painlevé equations (I-V kind) are degenerate cases of that of the VI kind, in which some poles degenerate in multiple poles of the matrix $A(x)$.

The Riemann problem has a field theoretical interpretation. Assume for simplicity that all pole points are on the real line. Take a branch cut as in fig.2.2:

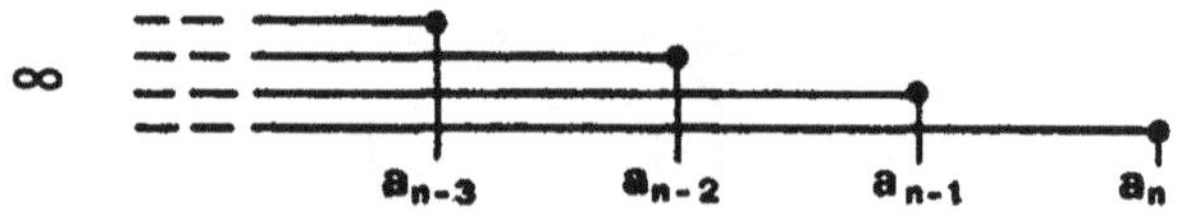

fig. 2.2

Suppose one has the solution to the Riemann problem $Y(x)$, and let $Y_+(x)$ and $Y_-(x)$ denote its branches defined in the upper and lower complex plane respectively. The two functions $Y_\pm(x)$ in the two semi-planes are connected, via analytical continuation by matrices M_γ: for examples $Y_+(x) = Y_-(x)M_n$ if the path goes between a_{n-1} and a_n, $Y_+(x) = Y_-(x)M_{n-1}M_n$ if the path goes between a_{n-2} and a_{n-1} and so on. On the real axis the two matrices $Y_\pm(x)$ are related as follows:

$$Y_+(x) = Y_-(x)M(x) \tag{2.16a}$$

$$M(x) = M_1(x)M_2(x)\dots M_n(x) \tag{2.16b}$$

$$M_\gamma(x) = \begin{cases} M_\gamma & (x < a_\nu) \\ 1 & (a_\nu < x) \end{cases} \tag{2.16c}$$

Let us introduce free fermion operators on the real line $\psi^{(i)}(x), \psi^{*(i)}(x)$; $i = 1, \dots, m$

$$[\psi^{(i)}(x), \psi^{*(j)}(x')]_+ = \delta_{ij}\delta(x - x') \tag{2.17}$$

and consider the vacuum expectation value:

$$< 0 \mid \psi^{(i)}(x)\psi^{*(j)}(x') \mid 0 >= \delta_{ij}\frac{1}{2\pi i}\frac{1}{x - x' + io} \tag{2.18}$$

Suppose that a field operator φ exists which satisfies the following defining commutation relations:

$$\psi^{(j)}(x)\varphi = \sum_{i=1}^{m} \varphi\psi^{(i)}(x)\big(M(x)\big)_{ij} \tag{2.19a}$$

$$\varphi\psi^{*(j)}(x) = \sum_{i=1}^{m} \psi^{*(i)}(x)\varphi\big(M(x)\big)_{ji} \tag{2.19b}$$

where the matrix $M(x)$ is that of (2.16 b). The solution of the Riemann problem can now be given in terms of φ and the free fermion operators as follows:

$$Y_+(x) = -2\pi i(x-x_0)\frac{<0\mid \psi^{(j)}(x)\psi^{*(i)}(x_0)\varphi\mid 0>}{<0\mid\varphi\mid 0>} \tag{2.20a}$$

$$Y_-(x) = 2\pi i(x-x_0)\frac{<0\mid \psi^{*(i)}(x_0)\varphi\psi^{(j)}(x)\mid 0>}{<0\mid\varphi\mid 0>} \tag{2.20b}$$

where $Y_\pm(x_0) = 1$ From the defining relations (2.19) and the factorization (2.16 b), it follows that φ can also be given in a factored form:

$$\varphi = \varphi(a_1; L_1)\dots\varphi(a_n; L_n) \tag{2.21}$$

where each $\varphi(a_\nu; L_\nu)$ corresponds to the Riemann problem with only one branch point a_ν. Explicitly it reads

$$\varphi(a;L) = :\exp\left\{\iint \sum R_{ij}(x-a;x'-a;L)\psi^{(i)}(x)\psi^{*(j)}(x')dx\,dx'\right\}:$$

$$R_{ij}(x,x';L) = \frac{i}{\pi}\sin\pi L\cdot x_-^L x_-'^{-L}\left(\frac{i}{x-x'+io}e^{\pi iL} + \frac{-i}{x-x'-io}e^{-\pi iL}\right) \tag{2.22}$$

(:a: denote the normal ordered form of a).

One can check that (2.20) are solution of the Riemann problem provided $|L_\nu|$ are sufficiently small (this is done by expanding (2.22) in a series). We want to emphasize that the operator in eq. (2.22) — apart from normal ordering — is an exponential of a bilinear combination of free fermion operators. As commutators of quadratic products of free fermions (or bosons) are again quadratic, they constitute a Lie algebra: $\varphi(a;L)$ looks as a group element corresponding to such a Lie algebra. The matrices A_ν are now related to $Y(x)$ because it satisfies the linear equations (2.13): so the solutions of the Schlesinger equations are constructed as espectation values of group elements via (2.20–2.22). We can say that this is the "hidden symmetry group" of the Riemann problem.

3. ISING CORRELATION

Let us consider a two dimensional Ising model defined by the Hamiltonian:

$$E = \sum_{i,j} (-E_1 \sigma_{ij}\sigma_{i+1j} - E_2 \sigma_{ij}\sigma_{ij+1}) \tag{3.1}$$

where the spin variables σ_{ij} are dicotomic ($\sigma_{ij} = \pm 1$) and are given for every site (i, j) of a planar square lattice (see fig. 3.1)

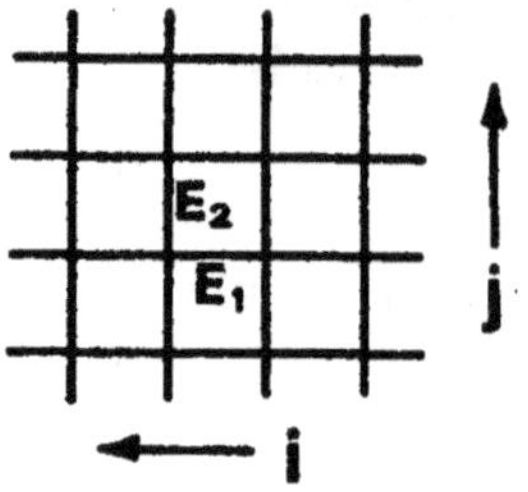

fig. 3.1

The partition function Z is the sum over all the spin configurations $\underline{\sigma}$ of the Boltzmann's weights:

$$Z = \sum_{\underline{\sigma}} e^{-\beta E(\underline{\sigma})} \tag{3.2}$$

The two point correlation function is obtained by averaging the product of two spin variables on two different sites over all spin configurations:

$$<\sigma_{00}\sigma_{MN}> = Z^{-1} \sum_{\underline{\sigma}} \sigma_{00}\sigma_{MN} e^{-\beta E(\underline{\sigma})} \tag{3.3}$$

The correlation function (3.3) can be evaluated using the row-to-row transfer matrix V (see Baxter's paper in this book):

$$\begin{aligned} <\sigma_{00}\sigma_{MN}> &= \frac{\mathrm{tr}(\hat{\sigma}_{00}\hat{\sigma}_{MN}V^N)}{\mathrm{tr}V^N} \qquad (N \geq 0) \\ \hat{\sigma}_{ij} &= V^j \hat{\sigma}_i^z V^{-j} \end{aligned} \tag{3.4}$$

In eqs. (3.4) $\hat{\sigma}_i^\alpha$ are spin operators (generalized Pauli matrices). We now convert the spin operators $\hat{\sigma}_i$ in free fermion operators by means of the Jordan-Wigner transform:

$$\begin{aligned} p_j &= \hat{\sigma}_1^x \hat{\sigma}_2^x \cdots \hat{\sigma}_{j-1}^x \hat{\sigma}_j^z \\ q_j &= \hat{\sigma}_1^x \hat{\sigma}_2^x \cdots \hat{\sigma}_{j-1}^x i\hat{\sigma}_j^y \end{aligned} \tag{3.5}$$

It is easy to see that p_j and q_j are free fermion operators, i.e. they anticommute for $i \neq j$ and $p_j^2 = 1$; $q_j^2 = -1$. It follows that if w and w' are any two elements of W, which is the linear span of p_j and q_j, then their anticommutator is a complex number:

$$[w, w']_+ = a \qquad a \in \mathbb{C}\,, \forall w, w' \in W \tag{3.6}$$

The free fermions $w \in W$ generate an associative algebra $\mathcal{A}(W)$ which is a Clifford algebra. The subspace of $\mathcal{A}(W)$ spanned by all the bilinear product of free fermions is a Lie algebra, as bilinear products close under commutation:

$$\mathcal{I}(W) = \left\{ \sum_{i,j} c_{ij} w_i w_j \mid w_i \in W, c_{ij} \in \mathbb{C} \right\} \tag{3.7}$$

The corresponding Lie group, called the Clifford group, is defined as follows:

$$\mathcal{G}(w) = \{ g \in \mathcal{A}(W) \mid \exists g^{-1},\, g w g^{-1} \in W \ \forall w \in W \} \tag{3.8}$$

(It is very easy to check that the Lie algebra of $\mathcal{G}(W)$ is exactly $\mathcal{I}(W)$).

In the Ising model both the transfer matrix V and the spin operators $\hat{\sigma}_j^z$ belong to the Clifford group $\mathcal{G}(W)$: this fact (which constitute the main result of Onsager's work in 1944 [6]) makes the model solvable. Solving eqs. (3.5) for the spin operators one obtains:

$$\hat{\sigma}_j^z = (p_1 q_1)(p_2 q_2) \cdots (p_{j-1} q_{j-1}) p_j \tag{3.9}$$

The spin operators $\hat{\sigma}_j^z$ thought of as elements of $\mathcal{G}(W)$ operate on W according to the relations:

$$\begin{aligned} \hat{\sigma}_j^z p_i (\hat{\sigma}_j^z)^{-1} &= \begin{cases} +p_i & (i \leq j) \\ -p_i & (i > j), \end{cases} \\ \hat{\sigma}_j^z q_i (\hat{\sigma}_j^z)^{-1} &= \begin{cases} +q_i & (i < j) \\ -q_i & (i \geq j). \end{cases} \end{aligned} \tag{3.10}$$

The relevance of the Clifford group will become evident in the following.

In general, suppose W is a vector space on $\mathbb{C}$ equipped with a non-degenerate bilinear form:

$$w \cdot w' + w' \cdot w = (w, w') \qquad \forall w, w' \in W \tag{3.11}$$

The Clifford algebra $\mathcal{A}(W)$ over W is defined by imposing the condition that the product of two elements $w, w' \in W$ is reduced to a number taking the anticommutator. So eq. (3.11) is the defining relation of the algebra $\mathcal{A}(W)$ (if the bilinear form is identically zero one obtains a so-called Grassmann algebra). The Clifford group is defined according to the previous scheme (see eq.(3.8)). The linear transformation of W by $\mathcal{G}(W)$ leaves the bilinear form (3.11) invariant:

$$\begin{aligned} (T_g(w), T_g(w')) &= (w, w') \\ T_g(w) &= gwg^{-1} \in W \end{aligned} \tag{3.12}$$

So T_g belongs to the ortogonal group $\mathcal{O}(W)$ of the vector space W, which is a subgroup of the complete linear transformation group $\mathcal{GL}(W)$ of W. Changing g by a constant factor $g = cg'$, $(c \neq 0)$ leaves T_g obviously unchanged $T_g = T_{g'}$. Conversely it can be shown that for $g, g' \in \mathcal{G}(W)$, $T_g = T_{g'}$ implies $g = cg'$, $(c \neq 0)$. So, all the informations are coded in the linear realization T_g of the group elements.

In table 3.1 we list the dimensions of the above spaces.

	W	$\mathcal{O}(W)$	$\mathcal{G}(W)$	$\mathcal{A}(W)$
dim	n	$\frac{1}{2}n(n-1)$	$\frac{1}{2}n(n-1)+1$	2^n

table 3.1

The dimension of $\mathcal{G}(W)$ is one more than that of $\mathcal{O}(W)$ (due to the invariance property of T_g which we mentioned before). The total Clifford algebra has a dimension which is much bigger than that of $\mathcal{G}(W)$: this makes $\mathcal{G}(W)$ very useful in the thermodynamic limit $(n \to \infty)$ where free fermions become fermion fields and $\mathcal{G}(W)$ is controlled by functions of two variables.

Going back to the Ising model we want to note that in particular the time evolution of p_j and q_j by V are linear, so their equation of motion is also that of free fields (we take the vertical direction as the time direction and call $V^j p_i V^{-j}$ the time evolution of p_j by V).

Let us now take the continuum limit of the Ising model at the critical temperature, i.e. $\epsilon = |T - T_c| \to 0$ while fixing $\epsilon i = x^1$ and $\epsilon j = \sqrt{-1}x^0$. With the above substitutions we obtain a Minkowsky field theory: the free fermions tend to the free two components Majorana fermion:

$$\left. \begin{matrix} V^j p_i V^{-j} \\ V^j q_i V^{-j} \end{matrix} \right\} \longrightarrow \psi(x) = \begin{pmatrix} \psi_+(x) \\ \psi_-(x) \end{pmatrix} \tag{3.13}$$

These are just the standard Lorentz invariant two dimensional fields which satisfy the Dirac equation:

$$(\not{\partial}+m)\psi(x)=0 \tag{3.14}$$

Let us consider also the limit of the spin operators from above and from below the critical temperature:

$$\hat{\sigma}^z_{ij} \longrightarrow \begin{cases} \sigma(x) & T\to T_c+0 \\ \mu(x) & T\to T_c-0 \end{cases} \tag{3.15}$$

(In some cases they are called "disorder variable" and "order variable" by Kadanoff et al.).

Now we want to characterize the correlation functions of these operators by using solutions to some non linear differential equations like those of Painlevé type: the simplest case is that of the two points correlation function which is written in terms of the Painlevé function of the 3rd kind as we have already seen in the introduction.

The commutation relations (3.10) in the continuum limit become:

$$\psi_\pm(x)\sigma(a)=\begin{cases} +\sigma(a)\psi_\pm(x) & (x^1<a^1) \\ -\sigma(a)\psi_\pm(x) & (x^1>a^1) \end{cases} \tag{3.16}$$

In eq. (3.16) x belongs to the space-like cone of a as in fig. 3.2

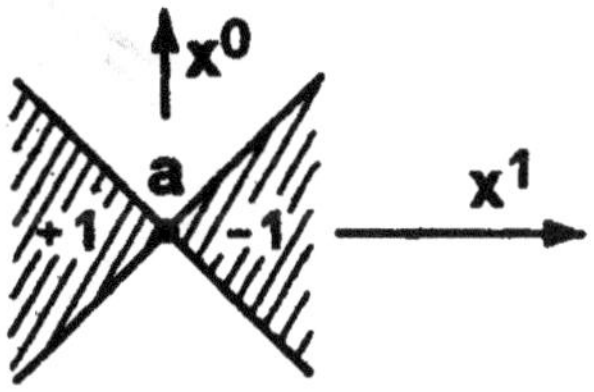

fig. 3.2

$\mu(a)$ satisfies commutation relations like (3.16) but with reversed sign. The $\mu(a)$'s can be written as normal ordered exponentials of quadratic forms of fermions:

$$\mu(a)=:e^{\rho(a)}: \qquad (\rho(a) \quad \text{quadratic} \quad \text{fermion}) \tag{3.17}$$

while $\sigma(a)$ contains a factor linear in free fermions:

$$\sigma(a)=:\psi_0(x)e^{\rho(a)}: \qquad (\psi_0(x) \quad \text{free} \quad \text{fermion}) \tag{3.18}$$

The explicit forms of $\rho(a)$ and $\psi_0(x)$ are given in terms of the Fourier transform $\hat{\psi}(u)$:

$$\psi_{\pm}(x) = \int \underline{du}\,(\sqrt{o+iu})^{\pm 1}\,\hat{\psi}(u) e^{-ipx} \tag{3.19}$$

$$p = \left(m\frac{u+u^{-1}}{2},\, m\frac{u-u^{-1}}{2} \right) \qquad -\infty < u < \infty$$

$$\underline{du} = \frac{du}{2\pi u}$$

One has precisely:

$$\psi_0(x) = \int \underline{du}\,\hat{\psi}(u) e^{-ipx} \tag{3.20}$$

$$\rho(x) = \iint \underline{du du'}\,\frac{-i(u-u')}{u+u'-io}\,\hat{\psi}(u)\hat{\psi}(u') e^{-i(p+p')x}$$

Eq. (3.20) can also be viewed as the definitions of the fields $\psi_0(x)$ and $\rho(x)$. The variable labels the momentum on the mass shell in the way shown in fig. 3.3.

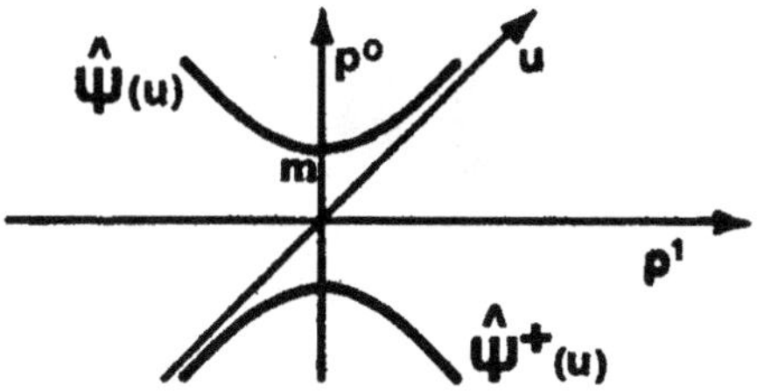

fig. 3.3

For $u > 0$ $\hat{\psi}(u)$ is an annihilation operator, $\hat{\psi}(-u) = \hat{\psi}^+(u)$, $u > 0$ a creation operator. We want now to compute vacuum expectation values of products of fields. The simplest case is that of only one field; we compute the following so called "wave functions":

$$\begin{aligned} w_{\pm}(x) &= < 0 \mid \psi_{\pm}(x)\sigma(a) \mid 0 > \\ w'_{\pm}(x) &= < 0 \mid \sigma(a)\psi_{\pm}(x) \mid 0 > \end{aligned} \tag{3.21}$$

The wave functions (3.21) are living in the Minkowsky space: we consider also the Euclidean continuation (see fig. 3.4). Due to the annihilation or creation character of $\psi_{\pm}$ as a function of the frequency the analytical continuation can be done as in fig. 3.4:

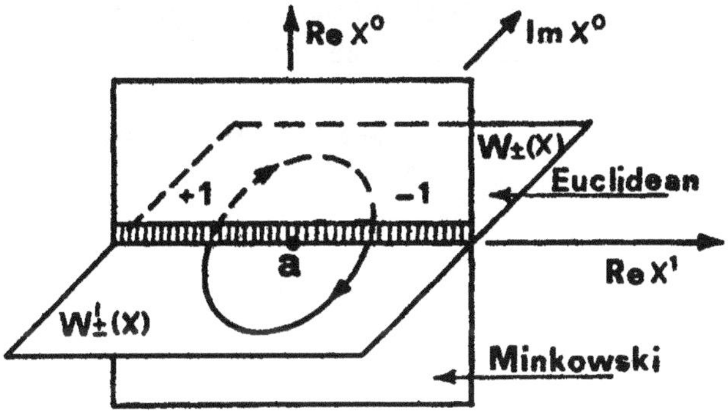

fig. 3.4

Consider the analytical continuation around the point a of the wave functions: we start on the positive x^1 axis considering $w'_{\pm}(x)$ and going around in the clockwise direction like in fig. 3.4. On the negative real x^1 axis $w_{\pm}$ and $w'_{\pm}$ are equal, due to the commutation in (3.16): so $w'_{\pm}(x)$ can be analytically prolonged to $w_{\pm}(x)$ and then we go back to $-w'_{\pm}(x)$ because of the anticommutation relation in (3.16). As a result going around a along the path in fig. 3.4 we get a minus sign: the wave functions are not single valued but they change sign going around a; the commutation relations imply the monodromy property for $w_{\pm}(x)$. The particular case of only one field as in (3.21) is very simple: one can give the explicit form for the wave function ($a = 0$):

$$w(x) = \begin{pmatrix} \frac{1}{\sqrt{z}} e^{-2m|z|} \\ \frac{1}{\sqrt{\bar{z}}} e^{-2m|z|} \end{pmatrix} \tag{3.22}$$

$$z = \frac{x_1 + ix_2}{2} \qquad x_2 = ix^0$$

In (3.22) the monodromy property is due to the square root: it is not an holomorphic function as it depends on $z, \bar{z}$ and $|z|$ and this is not completely analogus to the standard Riemann-Schlesinger problem.

In the general case we have to compute wave functions of the form:

$$W_{\pm}(x) = < 0 \mid \psi_{\pm}(x)\mu(a_1) \cdots \sigma(a_j) \cdots \mu(a_n) \mid 0 > \tag{3.23}$$

These functions have the following properties:

(i) as $\psi_\pm$ satisfy the Minkowski-Dirac equation, W must satisfy the Euclidean version of the Dirac equation:

$$(\underset{E}{\not\partial} + m)W = 0 \qquad z \neq a_1, \ldots, a_n \tag{3.24}$$

(ii)

$$W \sim e^{-2m|z|} \qquad |z| \to \infty \tag{3.25}$$

(iii)

$$W \sim (z - a_\nu)^{-\frac{1}{2}} \qquad z \to a_\nu \tag{3.26}$$

Eq. (3.26) implies that W changes sign when continued around a_ν. It is possible to show that the properties (i)– (iii) almost characterize the W functions and there are exactly n linearly independent such functions (n being the number of points a_ν).

Now we want to characterize these wave functions in the context of linear differential equations. We remember that W satisfy the Euclidean Dirac equation (3.24). Let us look for an operator which commute with the Dirac operator:

$$\underset{E}{\not\partial} = \gamma\partial + \overline{\gamma}\overline{\partial} = \begin{pmatrix} 0 & \overline{\partial} \\ \partial & 0 \end{pmatrix} \tag{3.27}$$

Besides the trivial ones ∂ and $\overline{\partial}$, one has the rotation operator:

$$M_F = z\partial - \overline{z}\overline{\partial} + \frac{1}{2}\gamma_5 \qquad \gamma_5 = \begin{pmatrix} 1 & 0 \\ 0 & -1 \end{pmatrix} \tag{3.28}$$

From equation (3.22) (with $z, \overline{z}$ replaced by $z - a$, $\overline{z} - \overline{a}$) one sees easily in the case of $n = 1$:

$$M_F w = (a\partial - \overline{a}\overline{\partial})w \tag{3.29}$$

In the general case, suppose that $W_1, \ldots, W_n$ constitute a basis for the wave functions, and consider the wave function:

$$M_F W_i = \sum_j (b_{ij}\partial W_j - \overline{b}_{ij}\overline{\partial} W_j + e_{ij} W_j) \tag{3.30a}$$

or in matrix notation:

$$M_F \mathbf{W} = (\mathbf{B}\partial - \overline{\mathbf{B}\partial} + \mathbf{E})\mathbf{W} \tag{3.30b}$$

Here $\mathbf{B}$ (resp. $\overline{\mathbf{B}}$) has $a_1, \ldots, a_n$ (resp. $\overline{a}_1, \ldots, \overline{a}_n$) as its eigenvalues, generalizing eq.(3.29). Because M_F commute with the Dirac operator the functions (3.30) also satisfy the Dirac equation and due to the form (3.30) they also behave exponentially at infinity and the monodromy properties must not change. The only change is the power of the growth (3.26) which is now $(z - a_\nu)^{-\frac{3}{2}}$ due to the differentiations in (3.30): but one can subtract this leading terms by a suitable definition of $\mathbf{B}$ and $\overline{\mathbf{B}}$ getting a function which grows moderately at a_ν.

Eq. (3.30) is a linear condition on the n linearly independent wave functions W_i: the Dirac equation gives only a very weak constraint as it means only a functional dependence of $\mathbf{W}$. On the contrary the eq. (3.30) along with the Dirac equation imply finite dimensionality of the space of solutions: so it gives the characterization of wave functions in terms of linear differential equations.

Still further we can also derive the wave functions with respect to a_ν:

$$\frac{\partial}{\partial a_\nu} W_i = \sum_j \Big(\Phi_{ij}^{(\nu)} \partial W_j - \overline{\Phi}_{ij}^{(\nu)} \overline{\partial} W_j + \Psi_{ij}^{(\nu)} W_j \Big) \tag{3.31}$$

(and similarly for the $\overline{a}_\nu$ derivative).

All we have said of eq. (3.30) is still true for eq. (3.31). If now one computes the integrability conditions for all equations (3.30) and (3.31) one gets NLPDE for all the coefficients $b_{ij} \ldots \Psi_{ij}^{(\nu)}$ (they are functions of the a_ν). For $n = 2$, the Painlevé equation of the 3rd kind is obtained.

In the limit $m \to 0$ everything become very simple: either the z or $\overline{z}$ dependence drop off and all one gets is just the usual functions in one complex variable. In this case the monodromy is just abelian (there is only a sign change and one can write the solution in terms of square roots etc.).

We can now connect these non linear equations to the correlation functions using the operators expansions of the form:

$$\begin{aligned} i\psi(x)\sigma(a) = {} & \frac{1}{2}\mu(a)\big(W_0[a] - W_0^*[a]\big) + \\ & + \frac{1}{m}\Big(-\frac{\partial}{\partial a^-}\mu(a) W_1[a] - \frac{\partial}{\partial a^+}\mu(a) W_1^*[a] \Big) + \cdots \end{aligned} \tag{3.32a}$$

$$\begin{aligned} \psi(x)\mu(a) = {} & -\frac{1}{2}\sigma(a)\big(W_0[a] + W_0^*[a]\big) + \\ & + \frac{1}{m}\Big(-\frac{\partial}{\partial a^-}\sigma(a) W_1[a] + \frac{\partial}{\partial a^+}\sigma(a) W_1^*[a] \Big) + \cdots \end{aligned} \tag{3.32b}$$

$$a^{\pm} = \frac{1}{2}(a^0 \pm a^1) \qquad \psi(x) = \begin{pmatrix} \psi_+(x) \\ \psi_-(x) \end{pmatrix} \tag{3.32c}$$

The expansions (3.32) are considered in the limit $x \to a$: $\Psi(x)$ satisfyes the Dirac equations and the expansions are written in terms of solutions of the Dirac equation:

$$W_j[a], W_j^*[a] \qquad (\not{\partial} + m)\mathbf{W} = 0 \tag{3.33}$$

which behave like:

$$W_j[a] \sim \begin{pmatrix} (x^- - a^-)^{j-\frac{1}{2}} \\ (x^- - a^-)^{j+\frac{1}{2}} \end{pmatrix} \tag{3.34a}$$

$$W_j^*[a] \sim \begin{pmatrix} (x^+ - a^+)^{j+\frac{1}{2}} \\ (x^+ - a^+)^{j-\frac{1}{2}} \end{pmatrix} \tag{3.34b}$$

Now we want to relate the correlation functions to the solutions of the differential equations.

More precisely let us define:

$$V_j = \frac{< 0 \mid \psi(x)\mu(a_1)\cdots\sigma(a_j)\cdots\mu(a_n) \mid 0 >}{< 0 \mid \mu(a_1)\cdots\mu(a_j)\cdots\mu(a_n) \mid 0 >} \tag{3.35}$$

We can compute the behavior at $x = a_i \quad (i \neq j)$ by using (3.32 b); we obtain:

$$V_j \sim \frac{\tau^{ij}}{\tau} \cdot -\frac{1}{2}(W_0[a_i] + W_0^*[a_i]) + \cdots \tag{3.36}$$

$$\tau = < 0 \mid \mu(a_1)\cdots\mu(a_n) \mid 0 > \tag{3.37a}$$

$$\tau^{ij} = < 0 \mid \mu(a_1)\cdots\sigma(a_i)\cdots\sigma(a_j)\cdots\mu(a_n) \mid 0 > \tag{3.37b}$$

$$\mathrm{Pf}\left(\frac{\tau^{ij}}{\tau}\right) = \frac{< 0 \mid \sigma(a_1)\cdots\sigma(a_n) \mid 0 >}{\tau} \tag{3.37c}$$

At $x = a_j$ using (3.32 a) we obtain:

$$V_j \sim \frac{1}{2}(W_0[a] - W_0^*[a]) + \frac{1}{2}\left(-\frac{\partial}{\partial a^-}\ln\tau \cdot W_1[a_j] - \frac{\partial}{\partial a^+}\ln\tau \cdot W_1^*[a_j]\right) \tag{3.38}$$

We have already seen that the functions (3.35) satisfy the LDE (3.30), (3.31) and that the integrability conditions for these equations turn to be NLPDE for all the coefficients $b_{ij} \cdots \Psi_{ij}^{(\nu)}$ the above expansions put in the LDE relate these coefficients to τ and τ^{ij} which can so be written in terms of solutions of NLPDE. In the general case these deformation equations are matrix non linear equations.

In (3.30 b) let us choose such a basis WWj that $\mathbf{B}$ becomes the diagonal matrix $\mathbf{A} = \text{diag}\{a_i\}$, and put $\overline{\mathbf{B}} = \mathbf{G}\overline{\mathbf{A}}\mathbf{G}^{-1}$, $\mathbf{E} = \mathbf{F}$. Then one has:

$$\begin{cases} d\mathbf{F} = -[\mathbf{F}, \Theta] + [d\mathbf{A}, {}^t\mathbf{G}\overline{\mathbf{A}}\mathbf{G}] + [\mathbf{A}, {}^t\mathbf{G}d\overline{\mathbf{A}}\mathbf{G}] \\ d\mathbf{G} = -\mathbf{G}\Theta + \overline{\Theta}\mathbf{G} \end{cases} \tag{3.39}$$

$$\mathbf{G} = (G_{ij}) \qquad \mathbf{F} = (F_{ij}) \qquad \Theta = (\Theta_{ij})$$

$$\Theta_{ij} = -F_{ij}\frac{d(a_i - a_j)}{a_i - a_j}$$

The ratios τ^{ij}/τ are related to $\mathbf{G} = e^{-2H}$ as

$$(\tanh \mathbf{H})_{ij} = \frac{\tau^{ij}}{\tau} \tag{3.40}$$

The logarithmic derivative of τ itself can be obtained as:

$$\omega = d\left(\frac{\ln \tau}{\sqrt{\det \cosh \mathbf{H}}}\right) \tag{3.41}$$

where ω is a 1-form given by:

$$\omega = -\frac{1}{2}\text{tr}(\mathbf{F}\Theta + \overline{\Theta}\mathbf{G}\mathbf{F}\mathbf{G}^{-1}) + \text{tr}[d(\mathbf{A}\overline{\mathbf{A}}) - \mathbf{G}^{-1}\overline{\mathbf{A}}\mathbf{G}d\mathbf{A} - \mathbf{G}\mathbf{A}\mathbf{G}{-}1d\overline{\mathbf{A}}] \tag{3.42}$$

Of course, the correlation functions correspond to one specific solution of the deformation equations (3.39). However, if we define ω by (3.42), then we can show that for any solution of the deformation equations:

$$d\omega = 0 \tag{3.43}$$

which means that ω, at least locally, can be written as:

$$\omega = d \ln \tau \tag{3.44}$$

with some function τ (not to be confused with the correlation function above), we can define the "τ-function" for any solution of the deformation equations. The situation is quite the same for the Schlesinger equations; let $\mathbf{A}_\nu$ be any solutions of them. If we define:

$$\omega = \sum_{\mu<\nu} \text{tr}\mathbf{A}_\mu\mathbf{A}_\nu d\ln(a_\mu - a_\nu) \tag{3.45}$$

we can show that equation (3.43) still holds.

In general we can introduce a τ function defined by (3.44).

In fact one can show that:

$$\tau = < 0 \mid \varphi(a_1; L_1) \cdots \varphi(a_n; L_n) \mid 0 > \qquad (3.46)$$

in the construction of the first section.

In conclusion we remark that the correlation function for the Ising model is just the τ function of the deformation equations.

We mention, at the end, that McCoy, Perk and Wu [7], using a more direct approach, construct non linear difference equations for the correlation functions on the lattice, which in the continuum limit looks like the above said NLPDE. Their result has two characteristic features:

(i) Most of the equations are also true for totally inhomogeneous lattice,

(ii) The NL equations are written in a special bilinear form (called Hirota's bilinear form).

Their origin will become apparent in sect. 5.

4. IMPENETRABLE BOSONS AND RELATED TOPICS

Another application of the monodromy idea to a physical problem is the impenetrable Bose gas. Here also the central point is the computation of the N-point correlation function using some type of NLDE (which includes the Painlevé equation of the V kind).

We start with the non linear Schrödinger equation in one dimension:

$$i\frac{\partial}{\partial t}\phi = -\frac{1}{2}\frac{\partial^2}{\partial x^2}\phi + c\phi^*\phi^2 \tag{4.1}$$

Eq. (4.1), as a classical equation, is a very typical example of soliton equation: on the contrary we will discuss this equation as a quantum one.

We consider ϕ as quantum field operators satisfying the equal time commutation relations:

$$\left[\phi(x,t),\phi^*(x',t)\right] = \delta(x-x') \tag{4.2}$$

As is well known, in the quantum mechanical language this is equivalent to consider an N-body problem. Let us consider a finite box $0 < x < L$ (the problem is a non relativistic one). The eq. (4.1) must be reviewed in the second quantization description, corresponding to a first quantization problem with an Hamiltonian:

$$H_N = -\frac{1}{2}\sum_{i=1}^{N}\frac{\partial^2}{\partial x_i^2} + c\sum_{i<j}\delta(x_i - x_j) \tag{4.3}$$

which is an N-body Hamiltonian with a delta-function two-body potential. The exact meaning of this Hamiltonian is that one has an eigenvalue problem:

$$-\frac{1}{2}\sum_{i=1}^{N}\frac{\partial^2}{\partial x_i^2}\psi_N = E\psi_N \tag{4.4}$$

supplemented by the boundary conditions:

$$\psi_N\Big|_{x_i=x_j+0} = \psi_N\Big|_{x_i=x_j-0} \tag{4.5}$$

$$\left(\frac{\partial\psi_N}{\partial x_i} - \frac{\partial\psi_N}{\partial x_j}\right)\Bigg|_{x_i=x_j+0} - \left(\frac{\partial\psi_N}{\partial x_i} - \frac{\partial\psi_N}{\partial x_j}\right)\Bigg|_{x_i=x_j-0} = 2c\psi_N\Big|_{x_i=x_j\pm 0}$$

and we must assume that $\psi_N(x_1,\ldots,x_N)$ is symmetric.

Thê equation (4.1) is so almost a free equation and one can expect that inside the box the solution has the exponential form:

$$\psi_N(x_1,\ldots,x_N;k_1,\ldots,k_N;c) = Z_N^{-1} \sum_{P\in S_N} A_P(k;c)e^{ik_{P(1)}x_1+\cdots+ik_{P(N)}x_N}$$
$$\forall x_i: \quad 0\le x_1 \le \cdots \le x_N \le L \tag{4.6}$$

and in fact Lieb and Liniger [8] showed that eq. (4.6) gives actually a solution for arbitrary N: Z_N is a normalization constant, $k_1,\ldots,k_N$ are costant momenta and the sum is over all the permutations P with coefficients $A_P(k;c)$ given by:

$$A_P(k;c) = \prod_{j<j'}\left[1-\frac{1}{ic}(k_{P(j)}-k_{P(j')})\right] \tag{4.7}$$

In order that the boundary conditions be satisfied it is necessary and sufficient to have coefficients of the form (4.7) up to some proportionality constant and the k_i values are limited by the condition:

$$(-)^{N-1}e^{ik_jL} = \prod_{j'(\neq j)} \frac{1-\frac{1}{ic}(k_j-k_{j'})}{1+\frac{1}{ic}(k_j-k_{j'})} \tag{4.8}$$

taking the boundary conditions as periodic too.

It is also known that for repulsive interaction, namely $c>0$, the possible choices of $k=(k_1,\ldots,k_N)$ give the complete set of eigenvectors for this problem. Lieb and Liniger constructed the above eigenfunctions by means of the well known so-called Bethe Ansatz.

We will not study the general case, but only the very special case when $c=+\infty$. In this case the boundary conditions (4.5) reduce to only one equation:

$$\psi_N\Big|_{x_i=x_j} = 0 \tag{4.9}$$

We have now a system of bosons which cannot cross each other, i.e. they are impenetrable. In this case the wave function reduces drastically to the very simple form:

$$\psi_N = |\psi_{N,FF}|$$
$$\psi_{N,FF} = \frac{1}{\sqrt{N/L^N}}\det\left(e^{ik_jx_{j'}}\right) \tag{4.10}$$

namely it is just an absolute value of a determinant; the condition (4.9) is obviously satisfyed by the determinantal form, which is actually antisymmetric (i.e. a wave function for free fermions) but turns to be symmetric taking the absolute value of it. In some sense the problem reduces to the free fermion problem. Actually this is more apparent in the language of quantum inverse method (which have been invented by several groups: Fadeev et al. [9], Thacker [10], Honerkamp et al. [11]). As already mentioned, the classical non linear Schrödinger equation is a soliton one and perhaps one of the first examples which have been solved by means of the inverse scattering technique. This technique relates the problem to a scattering problem on the real line: the reflection coefficient, defined in the usual way, becames a functional of the fields $r = r(k; \phi, \phi^*)$ Using the Gelfand-Levitan [12] equation one can recover the ϕ fields in terms of r. In the quantum case one follows exactly the same method: the r are now operators which can be formally written in terms of series involving the fields ϕ, ϕ^*. One can show that r satisfy very simple commutation relation which in our case (for any c) can be written as follows:

$$r(k)r(k') = \left(-\frac{1 - \frac{1}{ic}(k-k')}{1 + \frac{1}{ic}(k-k')}\right) r(k')r(k) \tag{4.11}$$

(In the classical case the time evolutions of the fields is not linear, while that of the reflection coefficients is just linear: this make the method very useful). When $c = +\infty$ the commutation relations (4.11) reduces to:

$$\left[r(k), r(k')\right]_+ = 0 \tag{4.12}$$

i.e. $r(k)$ is a free fermion and in fact the formula to recover ϕ and ϕ^* from r reduces to a sort of Jordan-Wigner transformation, which means that $\phi(x)$ belongs to the Clifford group [13].

So one could expect that in this example also the monodromy technique can be used.

The problem is to compute the n-particle reduced density matrix:

$$\rho_n(x_1, \ldots x_n; x'_1, \ldots, x'_n; c) = \lim_{N,L\to\infty} < \psi^0_{NL} \mid \phi^*(x_1, 0) \cdots \phi^*(x_n, 0)\phi(x'_1, 0) \cdots \phi(x'_n, 0) \mid \psi^0_{NL} >$$

$$N/L = \rho_0 : \quad \text{fixed} \tag{4.13}$$

i.e. a $2n$ point correlation function. ψ^0_{NL} represent the ground state wave function, ρ_0 turns to be the density of particles. Of course the problem is trivial if $\rho_0 = 0$, because in this case ψ^0_{NL} is just the vacuum and one obtains delta functions. The non trivial case is that of fixed finite ρ_0 (which in the relativistic quantum theory corresponds to treat the filled "Dirac sea"). The full result is known only for $c = +\infty$ (cf. Creamer-Thacker-Wilkinson, [12], also give an infinite series expression for finite c and $n = 2$).

In what follows $c = +\infty$ is assumed throughout.

There are two approaches to this problem.

The first one is due to Vaidya and Tracy [14]. One consider the XY Hamiltonian:

$$H = -\frac{1}{4}\sum_m [(1+\gamma)\sigma^x_m\sigma^x_{m+1} + (1-\gamma)\sigma^y_m\sigma^y_{m+1} + 2h\sigma^z_m] \tag{4.14}$$

and then relates this problem in the double scaling limit to the impenetrable Bose gas. γ and h are two parameters: the limit is done for $\gamma \to 0$ and $h \to 1$ fixing the ratio $\gamma/\sqrt{1-h^2} = g$ and then taking $g \to 0$. Vaidya and Tracy showed that by taking this double limit one gets the answer for the impenetrable Bose gas. It is also well known that the XY model is equivalent to a free fermion problem (much as the Ising model is) and so, by treating it in this way, it is from the beginning a free fermion problem.

The second approach is to use the Fredholm integral equation of the second kind: since the first approach is quite similar to the Ising case we will concentrate our attention on the second one.

Let us fix the notation. Suppose we have a Fredholm integral equation of this form:

$$u(x) - \lambda \int_I K(x,x')u(x')dx' = f(x) \tag{4.15}$$

(we obviously consider the one-dimensional case).

The Fredholm's determinant is defined, as usual, as follows:

$$\Delta_I(\lambda) = \sum_{l=0}^{\infty} \frac{(-\lambda)^l}{l!} \int_I \cdots \int_I dx_1 \cdots dx_l \, \mathrm{K}\binom{x_1,\ldots,x_l}{x_1,\ldots,x_l} \tag{4.16}$$

where we have used Fredholm's notation:

$$\mathrm{K}\binom{x_1,\ldots,x_l}{x'_1,\ldots,x'_l} = \det\left(k(x_j,x'_k)\right)_{j,k=1,\ldots,l} \tag{4.17}$$

We also write Fredholm's r^{th} minor determinant:

$$\Delta_I\begin{pmatrix} x_1,\ldots,x_r \\ x_1',\ldots,x_r' \end{pmatrix};\lambda\Big) =$$
$$\sum_{l=0}^{\infty} \frac{(-\lambda)^{l+r}}{l!} \int_I \cdots \int_I dx_{r+1}\cdots dx_{r+l}\, \mathrm{K}\begin{pmatrix} x_1 \ldots x_r & x_{r+1} \ldots x_{r+l} \\ x_1' \ldots x_r' & x_{r+1} \ldots x_{r+l} \end{pmatrix} \tag{4.18}$$

Now let:

$$k(x, x') = \frac{\sin(x - x')}{x - x'} \tag{4.19}$$

The Lenard [15] results states that the density matrix with coupling $c = +\infty$ can be written as a Fredholm's minor determinant of the form (with the kernel (4.19)):

$$\rho_n(x_1, \ldots, x_n; x_1', \ldots, x_n'; \infty) = \frac{\pm}{(-2)^n} \Delta_I\begin{pmatrix} x_1, \ldots, x_n \\ x_1', \ldots, x_n' \end{pmatrix}; \frac{2}{\pi}\Big) \tag{4.20}$$

$$\pm = \mathrm{sgn}\begin{pmatrix} x_1, \ldots, x_n \\ a_1, \ldots, a_n \end{pmatrix} \mathrm{sgn}\begin{pmatrix} x_1', \ldots, x_n' \\ a_{n+1}, \ldots, a_{2n} \end{pmatrix}$$

where $a_1 < \ldots < a_{2n}$ is a reordering of $x_1, \ldots, x_n, x_1', \ldots, x_n'$ and I is the union of closed intervals (see fig. 4.1):

$$I = [a_1, a_2] \bigcup \cdots \bigcup [a_{2n-1}, a_{2n}] \tag{4.21}$$

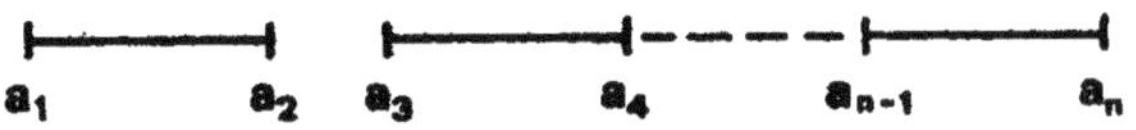

fig. 4.1

The eq. (4.20) results in the following way.

It is trivial to compute the free fermion density matrix, of course, and the result looks as follows:

$$\rho_{n,FF} = (-)^{n(n-1)/2} \det\left[\frac{\sin(x_j - x_k')}{\pi(x_j - x_k')}\right]_{j,k=1,\ldots,n} \tag{4.22}$$

Eq. (4.22) is only the first term of the Fredholm minor, by definition: namely it is the minor with $\lambda = 0$. The case $\lambda = 2/\pi$ corresponds, as already seen, to impenetrable bosons. (There is a third interesting case, that of $\lambda = 1/\pi$ which appears in the field of random matrices).

We will treat now the general case of λ and we will derive the NLDE satisfied by the Fredholm minors and determinants.

To obtain a NLDE first we will consider some linear differential equation and then study the monodromy structure and finally consider the deformation problem. So we first construct the solution of the LDE by means of the resolvent kernel. For the special case of $f(x) = K(x, x'')$ in eq. (4.15) the resolvent $u(x) \equiv R_I(x, x'')$ is obtained by means of the iterated integral equation:

$$R_I(x, x''; \lambda) = \sum_{l=o}^{\infty} \lambda^l \int_I \cdots \int_I dx_1 \ldots dx_l \, k(x_0, x_1) \cdots k(x_l, x_{l+1}) \quad (4.23)$$

$$(x_0 = x; x_{l+1} = x'')$$

and the Fredholm minor is related to this resolvent by the formula:

$$\Delta_I\binom{x_1, \ldots, x_n}{x'_1, \ldots, x'_n}; \lambda\Big) = (\,-\,)^n \Delta_I(\lambda) \det\left[R_I(x_j, x'_k; \lambda)\right] \quad (4.24)$$

Now we modify this definition slightly in order to get some analytic function on the complex plane. In eq. (4.23) x and x' are arbitrary complex numbers and the integral runs on the collection of intervals (4.21).

We change formula (4.23) by inserting the function:

$$h_I(x) = \frac{1}{2\pi i} \ln\left(\frac{x - a_1}{x - a_2} \cdots \frac{x - a_{2n-1}}{x - a_{2n}}\right) \quad (4.25)$$

between the kernels in eq. (4.23) and performing the integrals along the contour of fig. 4.2

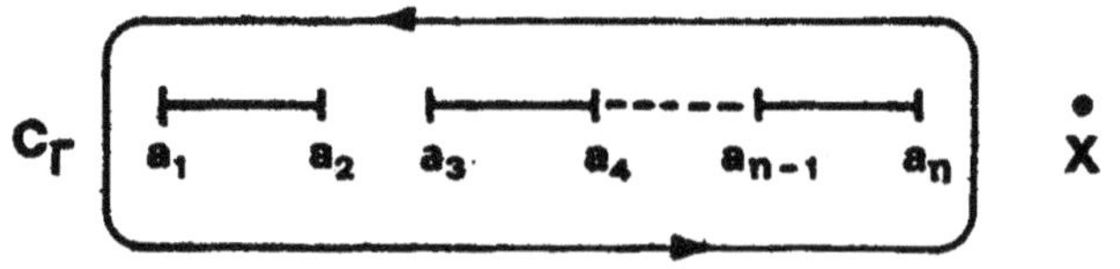

fig. 4.2

and putting x and x'' outside the contour C_I.

Next we let $x'' \to \infty$ and we extract a function only of x:

$$R_{\pm,I}(x;\lambda) = \sum_{l=o}^{\infty} \lambda^l \int_{C_I} \cdots \int_{C_I} dx_1 \cdots dx_l$$
$$\cdot e^{\pm i x_l} \frac{\sin(x_0 - x_1)}{x_0 - x_1} h_I(x_1) \frac{\sin(x_1 - x_2)}{x_1 - x_2} \cdots h_I(x_{l-1}) \frac{\sin(x_{l-1} - x_l)}{x_{l-1} - x_l} h_I(x_l) \tag{4.26}$$

Further we introduce also:

$$R^{\pm'}_{\pm,I}(x;\lambda) = \delta_{\pm',\pm} e^{\pm ix} + \sum_{l=1}^{\infty} \lambda^l \int_{C_I} \cdots \int_{C_I} dx_1 \cdots dx_l$$
$$\cdot \frac{\pm' e^{\pm' i(x_0 - x_1)}}{2i(x_0 - x_1)} h_I(x_1) \frac{\sin(x_1 - x_2)}{x_1 - x_2} \cdots h_I(x_{l-1}) \frac{\sin(x_{l-1} - x_l)}{x_{l-1} - x_l} h_I(x_l) e^{\pm i x_l} \tag{4.27}$$

Summing (4.27) on the upper indices we obtain:

$$R_{\pm,I}(x;\lambda) = R^{+}_{\pm,I}(x;\lambda) + R^{-}_{\pm,I}(x;\lambda) \tag{4.28}$$

In all the above formulas the point x is supposed to lie outside the contour C_I: the function $h_I(x)$ has singularities for x close to a_i and the function $\sin(x - x')/(x - x')$ is regular for x close to x'. In order to see the behavior near the a_i, we introduce again another function $\tilde{R}^{\pm'}_{\pm,I}(x;\lambda)$ which is the same as $R^{\pm'}_{\pm,I}(x;\lambda)$ but with an integration contour $\tilde{C}_I$ including x also:

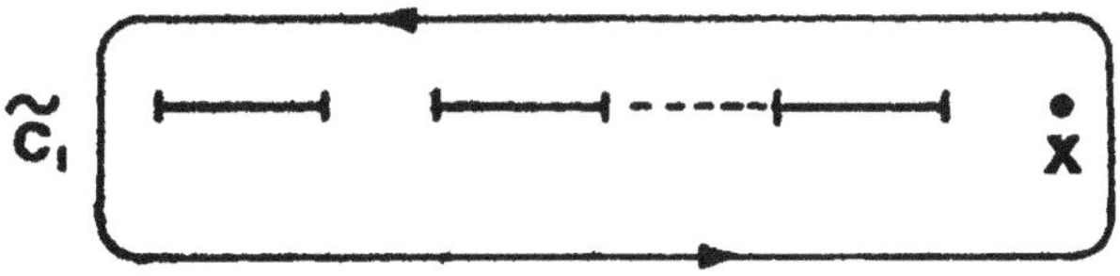

fig. 4.3

There hold the following relations between the above functions:

$$R^{\pm'}_{\pm,I}(x;\lambda) = \tilde{R}^{\pm'}_{\pm,I}(x;\lambda) \mp' R_{\pm,I}(x;\lambda) \frac{i\lambda}{2} \ln \frac{x - a_1}{x - a_2} \cdots \frac{x - a_{2n-1}}{x - a_{2n}} \tag{4.29}$$

$\tilde{R}^{\pm'}_{\pm,I}$ and $R_{\pm,I}$ are holomorphic functions at $x - a_i$ and the singularity structure of $R^{\pm'}_{\pm,I}$ is evident from (4.29). By means of the above function we define two-by-two matrix functions:

$$\mathbf{Y}(x) = \begin{pmatrix} R_{+,I}(x;\lambda) & R^{-}_{+,I}(x;\lambda) \\ R_{-,I}(x;\lambda) & R^{-}_{-,I}(x;\lambda) \end{pmatrix} \tag{4.30}$$

$$\mathbf{Y}_\infty(x) = \begin{pmatrix} R^{+}_{+,I}(x;\lambda) & R^{-}_{+,I}(x;\lambda) \\ R^{-}_{-,I}(x;\lambda) & R^{-}_{-,I}(x;\lambda) \end{pmatrix} \tag{4.31}$$

For x close to a_i the function (4.30) has the local decomposition:

$$\mathbf{Y}(x) = \hat{\mathbf{S}}(x)\left(\frac{x-a_1}{x-a_2}\cdots\frac{x-a_{2n-1}}{x-a_{2n}}\right)^{\mathbf{L}} \tag{4.32}$$

$$\mathbf{L} = \begin{pmatrix} 0 & \frac{i\lambda}{2} \\ 0 & 0 \end{pmatrix}$$

where $\hat{\mathbf{S}}(x)$ is holomorphic and invertible ($\det \hat{\mathbf{S}} = 1$) and the power must be understood as:

$$a^{\mathbf{L}} = e^{\mathbf{L}\ln a} = \begin{pmatrix} 1 & i\frac{\lambda}{2}\ln a \\ 0 & 1 \end{pmatrix}$$

So the singularity structure of $\mathbf{Y}$ is completely determined by (4.32) and we can also compute the monodromy around the points a_i (see fig. 4.4)

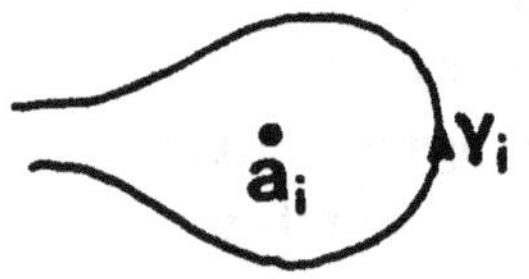

fig. 4.4

which is given by the matrix equation:

$$\mathbf{Y}(x) \xrightarrow{\gamma_i} \mathbf{Y}(x)\begin{pmatrix} 1 & \pm\pi\lambda \\ 0 & 1 \end{pmatrix} \quad \begin{pmatrix} + & i \text{ odd} \\ - & i \text{ even} \end{pmatrix} \tag{4.33}$$

When $x = \infty$ $\mathbf{Y}_\infty$ has a very simple structure:

$$\mathbf{Y}_\infty(x) = \left(\mathbf{1} + O\left(\frac{1}{x}\right)\right)\begin{pmatrix} e^{ix} & 0 \\ 0 & e^{-ix} \end{pmatrix} \tag{4.34}$$

The two matrices $\mathbf{Y}$ and $\mathbf{Y}_\infty$ are connected by the relation:

$$\mathbf{Y}(x) = \mathbf{Y}_\infty(x)\begin{pmatrix} 1 & 1 \\ 0 & 1 \end{pmatrix} \tag{4.35}$$

By an argument similar to that of Riemann in the usual case these properties lead to first order ODE with respect to x and PDE with respect to a_i:

$$\begin{cases} \dfrac{d}{dx}\mathbf{Y} = \left(\displaystyle\sum_{j=1}^{2n} \frac{\mathbf{A}_j}{x - a_j} + \mathbf{A}_\infty\right)\mathbf{Y} \\ \dfrac{\partial}{\partial a_j}\mathbf{Y} = -\dfrac{\mathbf{A}_j}{x - a_j}\mathbf{Y} \end{cases} \tag{4.36}$$

Eq. (4.36) looks very like the Riemann-Schlesinger equations except now there is an additional term $\mathbf{A}_\infty$ which accounts for the appearence of the exponential behavior at infinity.

Changing the variable at $x = \infty$ as $x = 1/t$ then the first eq. (4.36) becomes something like:

$$-\frac{d}{dt}\mathbf{Y} = \left(\frac{\mathbf{A}_\infty}{t^2} + \cdots\right)\mathbf{Y} \tag{4.37}$$

exhibiting a double pole at $t = 0$: this type of singularity is called "irregular singularity of rank 1". In the general theory of irregular singularity, one supposes to have singularities at $x = \infty$ for an equation of the form:

$$\frac{d\mathbf{Y}}{dx} = (\mathbf{A}_r x^{r-1} + \mathbf{A}_{r-1} x^{r-2} + \cdots)\mathbf{Y} \tag{4.38}$$

where the series between parenthesis is convergent at $x = \infty$. When $r < 0$, it is just a regular point. In the case $r = 0$ it is a regular singularity. When $r > 0$ it is called an irregular singularity of (Poincaré) rank k. In our case $r = 1$. Now suppose that the eigenvalues of the leading matrix $\mathbf{A}_r$ are distinct.

In this case one can find the formal series solution in the form:

$$\tilde{\mathbf{Y}}(x) = \mathbf{G}(1 + \mathbf{Y}_1 x^{-1} + \cdots)x^{\mathbf{D}} \exp(\mathbf{T}_1 x + \cdots + \mathbf{T}_r x^r) \tag{4.39}$$

where $\mathbf{D}, \mathbf{T}_1, \ldots, \mathbf{T}_r$ are diagonal matrices and $\mathbf{G}$, $\mathbf{Y}_i$ are some matrices.

The serie in (4.39) is not always convergent if $r \geq 1$, but it does represent the solution in the following sense. At the point $x = \infty$, if there is a singularity of rank k there exist $2r$ number of sectors surrounding this point (in fig. 4.5 $r = 2$)

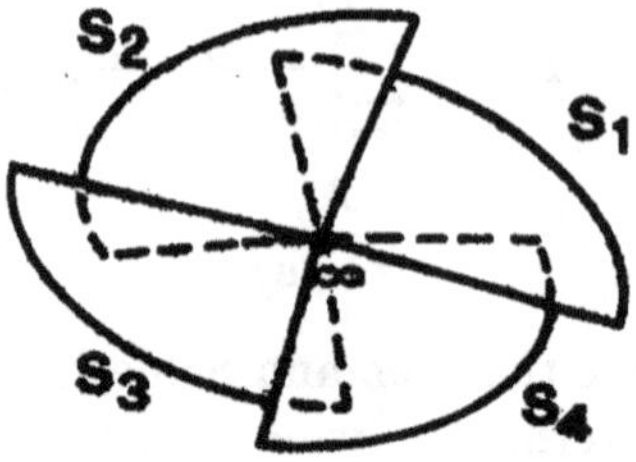

fig. 4.5

In each of the sectors S_j there exist one and only one solution to the LDE such that it has the serie (4.39) as asymptotic expansion. Now we have $2r$ different solutions $\mathbf{Y}_j(x)$ in each subsector and they may not be identical even if they have the same asymptotic expansion: this is called "Stokes phenomena". But since $\mathbf{Y}_j(x)$ are solution of the original equation (4.38) they should be related as follows:

$$\mathbf{Y}_{j+1}(x) = \mathbf{Y}_j(x)\mathbf{S}_j \tag{4.40}$$

where $\mathbf{S}_j$ are constant matrices called "Stokes multipliers". In this case, if the rank is r, then one has $\mathbf{Y}_{2r+1} = \mathbf{Y}_1\mathbf{S}_{2r}\mathbf{S}_{2r-1}\cdots\mathbf{S}_1$ by surrounding the point $x = \infty$, so the Stokes multipliers constitute a refined notion of monodromy in the case of $r > 0$. Now the generalized Riemann problem is stated in the following way. We give arbitrary points a_ν in the complex plane (possibly including infinity) and at each of these branch points the rank r_γ is fixed arbitrarly too. We have at each point the notion of Stokes multipliers (if the singularity is regular one has the notion of exponent of the monodromy) and also we have connections between the a_ν points (the theory is purely local), as for example eq. (4.35), which connect $x = \infty$ to finite points. Given these data can we find a differential equation corresponding to the monodromy data themselves? This is the generalized Riemann problem. The problem can be solved, at least in some region of the parameters space, and one can also consider the deformation equation of this problem.

Now the deformation parameters are not only the positions $a_1, \ldots, a_n$ but also the matrices in eq. (4.39) $\mathbf{T}_1^{(\nu)}, \ldots, \mathbf{T}_{r_\nu}^{(\nu)}$ $(1 \leq \nu \leq n,\ \nu = \infty)$ Now let us return to the case (4.36) of the impenetrable Bose gas: the resulting deformation equations have an hamiltonian form. Let us consider $2n$ τ-functions $(\tau_j, \tau_{-j}\quad j = 1, \ldots, n)$ and introduce the Poisson brackets:

$$\{\tau_{+j},\tau_{+k}\}=\{\tau_{-j},\tau_{-k}\}=0 \qquad \{\tau_{+j},\tau_{-k}\}=\delta_{jk} \tag{4.41}$$

Then let us define the hamiltonian one-form:

$$\omega=-\frac{1}{4}\sum_{j<k}\begin{vmatrix}\tau_{+j} & \tau_{+k}\\ \tau_{-j} & \tau_{-k}\end{vmatrix}\frac{da_j-da_k}{a_j-a_k}+i\sum_{j<k}\tau_{+j}\tau_{-k}da_j \tag{4.42}$$

The deformation equation is then given by:

$$d\tau_{\pm j}=\{\tau_{\pm j},\omega\} \tag{4.43}$$

When $n=2$ the equation (4.43) reads in the very simple form:

$$\left(x\frac{d^2\sigma}{dx^2}\right)^2=-4\left(x\frac{d\sigma}{dx}-1-\sigma\right)\left(x\frac{d\sigma}{dx}+\left(\frac{d\sigma}{dx}\right)^2-\sigma\right) \tag{4.44}$$

$$\sigma(x)=x\frac{d}{dx}\ln\rho(x) \tag{4.45}$$

$$\rho_1(x-x';\infty)=\rho(x-x') \tag{4.46}$$

where the density matrix is now traslationally invariant and depends on one parameter only.

The eq. (4.44) is an equivalent form of the Painlevé equation of the $V-th$ kind.

Finally we remark that ω is related to the Fredholm determinant in the simple way:

$$\omega=d\ln\Delta_I(\lambda) \tag{4.47}$$

$$\Delta_I\left(\begin{matrix}x_1\ldots x_n\\ x'_1\ldots x'_n\end{matrix};\lambda\right)=(-)^n\Delta_I(\lambda)\det\left[R_I(x_j,x'_k);\lambda\right] \tag{4.48}$$

$$R_I(a_j,a_k)=\begin{cases}\frac{1}{2i(a_j-a_k)}\begin{vmatrix}\tau_{+j} & \tau_{+k}\\ \tau_{-j} & \tau_{-k}\end{vmatrix} & (j\neq k)\\ \frac{1}{4}\sum_{l(\neq k)}\frac{1}{(a_l-a_k)^2}\begin{vmatrix}\tau_{+l} & \tau_{+k}\\ \tau_{-l} & \tau_{-k}\end{vmatrix}^2+i\tau_{+k}\tau_{-k} & (j=k)\end{cases} \tag{4.49}$$

So the Fredholm determinant is by definition a τ function for the corresponding monodromy problem: it is also the expectation value of Clifford operators. These are actually the general features of the deformation problem: in the most general case the τ function can be shown to be the Fredholm determinant of an integral equation which arises in solving the Riemann problem. Also the Clifford operators view can be generalized to the most general case of arbitrary number of irregular singularities. In sect. 5 we shall see that the notion of τ function can be further extended to include soliton theory as well.

5. SOLITON EQUATIONS AND INFINITE DIMENSIONAL LIE ALGEBRA

5.1 Introduction

This section is based on a work of E. Date, M. Jimbo, M. Kashiwara and T. Miwa done in 1981 [16, 17, 18, 19], which is inspired by the work of M. Sato and Y. Sato [20, 21].

In the previous sections Painlevé and Schlesinger equations have been studied, which are ordinary (Painlevé) and partial (Schlesinger) non linear differential equations: both have finite dimensional solutions, i.e. they depend on a finite number of integration constants.

In this section we will study more general equations, usually called "soliton equations", which on the contrary, need boundary conditions or initial conditions in order to determine the solutions: such conditions are obviously infinite dimensional. Our aim is to connect soliton equations to the representation theory of infinite dimensional Lie algebras by means of the τ function. In this way we will establish an equivalence between the group orbit of the so called "highest weight vector" and, on the other hand, the solution space of the soliton equations, the first being the Lie algebraic view and the second the differential equation one: the infinite dimensionality of the Lie algebra bears an infinite dimensional group orbit and so an infinite dimensional solution space. The most famous soliton equation is the KdV equation: this will be related to $A_1^{(1)}$, the simplest Kac Moody Lie algebra. There is another, less famous but more fundamental soliton equation which is called the KP (Kadomtsev-Petviashvili) equation: this will be related to $\mathcal{GL}(\infty)$.

The situation can be compared to that of the previous sections where Painlevé business was related to Onsager business. As mentioned before, there is a relation between Painlevé equation and soliton equation: on the other hand the key point of the Onsager solution was the introduction of free fermions and Clifford group, that is nothing but the group corresponding to $\mathcal{GL}(\infty)$. In the Onsager case one deals with very particular elements of the Clifford group, i.e. spin operators or transfer matrices: now we treat more general elements of the Clifford group.

In sect. 5.2 we will first explain two methods of solving soliton equations: one is the linearization method and the other the bilinearization method. The first way needs a "wave function": the key point is to introduce a hierarchy of infinitely many equations (corresponding to infinitely many Lax pairs). The second way uses another dependent variable which is nothing but our τ function: one of the advantages of this method is that it is very easy to derive explicit formulas for N soliton solutions. Then we will introduce vertex operators, which transform an N soliton to an $N+1$ soliton solution in a very explicit way: this is the most elementary way to explain the relation between soliton equations and infinite dimensional Lie algebras. In other words, $\mathcal{GL}(\infty)$ and $A_1^{(1)}$ appear as the Lie algebras of infinitesimal transformations of soliton solutions.

Sect. 5.3 will be devoted to the sistematic theory, namely we will introduce a "bilinear identity" which is understood most easily in the case of quasi periodic solutions. We will show that it is equivalent to both the linearization and the bilinearization methods, and so unifies them into one single identity: this will be done using the relation between τ functions and wave functions. Then in sect. 5.4 we will construct solutions of the bilinear identity in terms of free fermions and $\mathcal{GL}(\infty)$ using the Fock representation (this is called, in physicist language, "boson-fermion" correspondence). So τ or wave functions will be expressed as expectation values (a very special case was presented in sect. 3 in the Ising case: here we will treat more general cases). We will also establish that the bilinear identity is nothing but the characterization of the above said group orbit. At the end of this sextion we will give some examples including some affine Lie algebras (which are particular Kac Moody algebras) and the corresponding soliton equations.

5.2 Kadomtsev-Petviashvili (K-P) hierarchy and vertex-operators

The most famous example of soliton equation is the KdV equation which is a NLPDE in the indipendent variables t and x:

$$\frac{\partial u}{\partial t}-\frac{3}{2}u\frac{\partial u}{\partial x}-\frac{1}{4}\frac{\partial^3 u}{\partial x^3}=0 \tag{5.2.1}$$

The so called Lax pair is the following system of two linear differential equations:

$$\begin{aligned}\left(-\frac{\partial^2}{\partial x^2}-u\right)w&=0\\ \left(\frac{\partial}{\partial t}-\frac{\partial^3}{\partial x^3}-\frac{3}{2}u\frac{\partial}{\partial x}-v\right)w&=0\end{aligned} \tag{5.2.2}$$

(v is related to u).

The variable u is transformed to a new dependent variable w which satisfies linear equations.

In order to connect soliton theory to group representation theory it is better to deal with the KP equation:

$$\frac{3}{4}\frac{\partial^2 u}{\partial y^2}-\frac{\partial}{\partial x}\left(\frac{\partial u}{\partial t}-\frac{3}{2}u\frac{\partial u}{\partial x}-\frac{1}{4}\frac{\partial^3 u}{\partial x^3}\right)=0 \tag{5.2.3}$$

The KP equation contains KdV: actually the KdV equation is obtained by dropping out the y dependence. The KP Lax pair is now given by:

$$\begin{aligned}\left(\frac{\partial}{\partial y}-\frac{\partial^2}{\partial x^2}-u\right)w&=0\\ \left(\frac{\partial}{\partial t}-\frac{\partial^3}{\partial x^3}-\frac{3}{2}u\frac{\partial}{\partial x}-v\right)w&=0\end{aligned} \tag{5.2.4}$$

The KP equation is the most fundamental soliton equation and almost all soliton equations can be derived from the KP equation in some way.

The relation between KP and KdV equations is well understood by considering the corresponding Lie groups as we will explain later.

The inverse scattering method is nothing but dealing with the scattering theory of eqs. (5.2.2) and (5.2.4): Novikov and Krichever [22] dealt with quasi periodic solutions and their method was substantially another linearization method. For our theory the meaning of linearization is somewhat different; namely, using linearization we can introduce, in a natural manner, a hierarchy.

The KP equation contains three independent variables $x_1 = x$, $x_2 = y$, $x_3 = t$. If one introduces infinitely many variables $x = (x_1, x_2, x_3, \ldots)$

one can obtain a hierarchy of infinitely many equations in the following general way.

Let us consider the w function depending on x and on a single parameter k:

$$w(x,k) = \left(1 + \frac{w_1(x)}{k} + \frac{w_2(x)}{k^2} + \cdots\right) e^{\xi(x,k)} \qquad (5.2.5)$$

where:

$$\xi(x,k) = \sum_{n=1}^{\infty} k^n x_n \qquad (5.2.6)$$

k is considered near the point at infinity. In eq. (5.2.6) the index n of x_n is consistent with the power of k in the formal series. Now we require $w(x,k)$ to satisfy the following LDE's:

$$\frac{\partial}{\partial x_n} w(x,k) = B_n\left(x, \frac{\partial}{\partial x_1}\right) w(x,k) \qquad (5.2.7)$$

which generalize eqs. (5.2.2) and (5.2.4). In eq. (5.2.7) $B_n(x, \frac{\partial}{\partial x_1})$ contains all the parameters x but the derivatives are only with respect to x_1:

$$B_n(x, \partial_1) = \partial_1^n + u_{n,n-2}\partial_1^{n-2} + \cdots + u_{n,0} \qquad (5.2.8)$$

$$\partial_1 = \frac{\partial}{\partial x_1}$$

The expression (5.2.8) for $B_n(x, \partial_1)$ does not contain ∂_1^{n-1} correspondingly to the normalization in (5.2.5). We require now compatibility conditions for the system (5.2.7) of LDE. We differentiate the nth equation in (5.2.7) with respect to x_m and then the mth with respect to x_n and then subtract one from the other. The resulting equation contains only x_1 differentiations:

$$\{(\partial_n B_m) - (\partial_m B_n) - [B_n, B_m]\} w = 0 \qquad (5.2.9)$$

So eq. (5.2.9) is an ODE and the solution space is finite dimensional. But w contain k as arbitrary parameter which implies that the solution should be infinite dimensional: so the full operator in front of w must be zero. These equations give infinitely many constraints on the coefficients $u_{n,p}$: after infinitely many integrations we relate all the $u_{n,p}$'s to a single parameter u, thus obtaining infinitely many NLPDE for such u with respect to infinitely many independent variables. The extension of the hierarchy does not lose nor get anything, i.e. we have introduced infinitely many variables but we have imposed infinitely many constraints and the solution space does not change.

Let us now consider the bilinearization scheme. Instead of u we introduce a new variable τ. We need some notation. We write a polynomial $D = (D_1, D_2, \ldots)$ in front of two functions of $x = (x_1, x_2, \ldots)$ separated by a dot, to mean the following:

$$P(D)f(x) \cdot g(x) = P(\partial_y)\big(f(x+y)g(x-y)\big)\Big|_{y=0} \tag{5.2.10}$$

For example, denoting by z a single variable, one obtains the Leibniz rule with the minus sign for the Hirota notation:

$$\begin{aligned} \partial_z f(z)g(z) &= \big(\partial_z f(z)\big)g(z) + f(z)\partial_z g(z) \\ D_z f(z) \cdot g(z) &= \big(\partial_z f(z)\big)g(z) - f(z)\partial_z g(z) \end{aligned} \tag{5.2.11}$$

Hirota introduces the new variable τ by the transform:

$$u = 2\frac{\partial^2}{\partial x_1^2} \ln \tau \tag{5.2.12}$$

One can recognize a similarity between the last formula and the formula for the τ-function in sect. 3.

KP equation can now be rewritten in the Hirota bilinear form:

$$(D_1^4 + 3D_2^2 - 4D_1D_3)\tau \cdot \tau = 0 \tag{5.2.13}$$

The introduction of τ by (5.2.12) is rather mysterious: it is remarkable that this transformation can be always done for solvable soliton-type equations (for example the Sine-Gordon equation, which is far from linear). The advantage of trasforming non linear equations into Hirota's form is the following. One can seek for a solution in the form:

$$\tau = 1 + f_1 + f_2 + \cdots \tag{5.2.14}$$

$\tau = 1$ is a solution to any Hirota type equation. With the Ansatz (5.2.14) one obtains linear equations by successively adding f_n. For example, for a Hirota equation of the form:

$$P(D)\tau \cdot \tau = 0 \tag{5.2.15}$$

one obtains for f_1 the linear equation:

$$P(\partial_x)f_1 = 0 \tag{5.2.16}$$

The constant coefficient linear equation (5.2.16) can be solved in terms of a sum of exponentials. If f_1 is a single exponential one can stop here, i.e. $\tau = 1 + f_1$ is a solution of Hirota equation without any further term. But if one uses two exponential terms, then they alone cannot solve the Hirota equation and one needs another term with the product of the exponentials with some coefficient. For any Hirota equation one can then choose appropriately such coefficients to obtain a solution. The situation is very different if one starts with three exponentials for f_1. Only for the solvable equations a solution is found in the form $1 + f_1 + f_2 + f_3$. Then one can go further, i.e. an N soliton solution can be obtained in this iterative way.

In the above said manner we can construct the general N soliton solution to KP, which reads as follows:

$$\tau_N = 1 + \sum_{n=1}^{N} \sum_{i \le j_1 \le \cdots \le j_n \le N} \prod_{j=j_1,\ldots,j_n} a_j \prod_{\substack{j,j'=j_1,\ldots,j_n \\ j<j'}} c_{jj'} \times$$
$$\times \exp\left\{ \sum_{j=j_1,\ldots,j_n} \left(\xi(x,p_j) - \xi(x,q_j)\right) \right\} \tag{5.2.17}$$

$$c_{jj'} = \frac{(p_j - p_{j'})(q_j - q_{j'})}{(p_j - q_{j'})(q_j - p_{j'})}$$

Let us turn now to the representation theory of infinite dimensional Lie algebras.

Lepowsky and Wilson have constructed the basic representation of $A_1^{(1)}$. $A_1^{(1)}$ is a Kac-Moody Lie algebra, that is a Lie algebra which is generated by a set of generators: $e_j, f_j, h_j \quad (j \in I: \quad I$ finite set) with the following defining relations:

$$\begin{aligned} [e_i, f_j] &= \delta_{ij} h_j \\ [h_i, e_j] &= a_{ij} e_j \\ [h_i, f_j] &= -a_{ij} f_j \\ [h_i, h_j] &= 0 \end{aligned} \tag{5.2.18}$$

$$\begin{aligned} (ad e_i)^{1-a_{ij}} e_j &= 0 \\ (ad f_i)^{1-a_{ij}} f_j &= 0 \end{aligned} \tag{5.2.19}$$

$(a_{ij})_{i,j \in I}$ is called the (generalized) Cartan matrix.

The simplest case of such an algebra is $A_1 \approx sl(2, \mathbb{C})$ which is finite dimensional: in this case $I = \{1\}$ and the Cartan matrix is 1×1 and is equal to 2: one has:

$$e_1 = \begin{pmatrix} 0 & 1 \\ 0 & 0 \end{pmatrix} \quad f_1 = \begin{pmatrix} 0 & 0 \\ 1 & 0 \end{pmatrix} \quad h_1 = \begin{pmatrix} 1 & 0 \\ 0 & -1 \end{pmatrix} \tag{5.2.20}$$

The definition of a Kac-Moody Lie algebra is just a generalization of sl(2, $\mathbb{C}$).

$A_1^{(1)}$ has the following Cartan matrix:

$$(a_{ij}) = \begin{pmatrix} 2 & -2 \\ -2 & 2 \end{pmatrix} \tag{5.2.21}$$

The Cartan matrix (5.2.21) is singular so that $A_1^{(1)}$ Lie algebra is infinite dimensional.

A more explicit way to say what is $A_1^{(1)}$ is the following:

$$A_1^1(1) \approx \mathrm{sl}(2, \mathbb{C}\,[k, k^{-1}]) \times \mathbb{C}\, C \tag{5.2.22}$$

i.e. it is the algebra of 2×2 matrices with elements which are Laurent polynomials in one variable k (and this makes it an infinite dimensional Lie algebra) plus some center C (one effectively needs one more dimension): there is not a direct sum in (5.2.22), but a suitable central extension.

The basic representation is generated by a "vacuum" vector $\mid 0 >$ such that:

$$\begin{aligned} e_i \mid 0 > &= 0 \\ h_i \mid 0 > &= \delta_{io} \mid 0 > \end{aligned} \tag{5.2.23}$$

i.e. $\mid 0 >$ is annihilated by the e_i operators and h_i act as constant multiples (it looks very similar to a Fock representation). The word "basic" corresponds to the particular choice of the eigenvalues of h_i of eq. (5.2.23). Lepowsky and Wilson [23] constructed a more explicit realization of this representation. They showed that $A_1^{(1)}$ contains a Heisenberg subalgebra, namely an algebra satisfying the usual commutation relations:

$$A_1^{(1)} \supset K = \sum_{j\ odd} \mathbb{C}\, H_j \tag{5.2.24}$$

$$[H_j, H_k] = j\delta_{j+k,0}\, C \tag{5.2.25}$$

It is natural to count H_j by their degree j and only odd numbers appear in Lepowsky-Wilson's construction. One can identify the operators H_j by differentiations or by moltiplications:

$$H_j = \begin{cases} \frac{\partial}{\partial x_j} & j > 0 \\ -jx_{-j} & j < 0 \end{cases} \tag{5.2.26}$$

For the Heisenberg algebra the representation space V is given by the polynomials in the variables $x_1, x_3, x_5, \ldots$:

$$V = \mathbb{C}\left[x_1, x_3, x_5, \ldots\right] \tag{5.2.27}$$

It is true that the basic representation is already irreducible with respect to the Heisenberg subalgebra $\mathcal{H}$, so that we can construct the complete representation of $A_1^{(1)}$ on this space (this is not true if one considers higher representations). We now want to find the general expression for the generators e_i, f_i, h_i acting on the particular space (5.2.27). Lepowsky and Wilson define a "vertex operator":

$$X(p) = e^{2(px_1 + p^3x_3 + \cdots)} e^{-2(\frac{1}{p}\frac{\partial}{\partial x_1} + \frac{1}{3p^3}\frac{\partial}{\partial x_3} + \cdots)} \tag{5.2.28}$$

The operator in (5.2.28) acts on the function space (5.2.27) like a shift operator followed by a multiplication by an exponential (one should remember the Taylor formula $e^{a\partial_x} f(x) = f(x+a)$. The basic representation of $A_1^{(1)}$ is now constructed in terms of the representation space (5.2.27) and the vertex operator (5.2.28). Notice the absence of the even parameters in the representation space (5.2.27): this is in correspondence to the fact that the KdV equation contains only parameters with odd indices (we should drop not only the x_2 dependence from KP, but also that on all even parameters). Moreover, if one considers the Heisenberg algebra, one can see that the action of $\frac{\partial}{\partial x_j}$ corresponds to the function shift $f(x+a)$, while the action of x_1 corresponds to multiplication by an exponential $e^{xa} f(x)$ (more precisely they are the infinitesimal generators of these transformations). Note that if:

$$P(D) f(x) \cdot g(x) = 0$$

then:

$$P(D)(f(x+a) \cdot g(x+a)) = 0$$

and:

$$P(D)\big(e^{ax} f(x) \cdot e^{ax} g(x)\big) = 0$$

In this sense Hirota's equations admits $\mathcal{H}$ as a transformation Lie algebra. The remarkable fact is that the vertex operators of Lepowsky and Wilson act on KdV τ functions as infinitesimal Bäcklund transformations, i.e.

if one applies exponentials of vertex operators to a solution of KdV one obtains another solution to KdV. For the KP equation the vertex operator has the form:

$$X(p,q) = e^{\xi(x,p)-\xi(x,q)}e^{-\xi(\tilde{\partial},p^{-1})+\xi(\tilde{\partial},q^{-1})} \tag{5.2.29}$$

where:

$$\tilde{\partial} = \left(\partial_1, \frac{\partial_2}{2}, \frac{\partial_3}{3}, \ldots\right) \tag{5.2.30}$$

The vertex operator (5.2.29) depends on two parameters p,q: requiring $p^2 = q^2$ the Lepowsky-Wilson vertex operator (5.2.28) for KdV is recovered. Again the KP vertex operator (5.2.29) acts on the solutions of KP equation as an infinitesimal Bäcklund transformation. More precisely if one applies the exponential of the vertex operator to an N-soliton solution one obtains an $N+1$ soliton solution, namely:

$$e^{aX(p,q)}\tau_N = \tau_{N+1} \tag{5.2.31}$$

So the general solution can be obtained in the simple way:

$$\tau_N = \exp\left\{\sum_{j=1}^{N} a_j X(p_j, q_j)\right\} \cdot 1 \tag{5.2.32}$$

where a_j,p_j and q_j are the same as in (5.2.17), i.e. an N-soliton solution can be obtained from the trivial one by applying vertex operators. This means that, if one can establish that the vertex operators and 1 actually span a Lie algebra (and this is true), the group orbit of the highest weight vector (which is 1, in this language) is equivalent to the space of solutions of soliton equations. In the next section we will explain how this special form of vertex operator enters the theory by treating the whole theory systematically.

5.3 Bilinear identities

First of all, we will unify the linearization and the bilinearization methods by means of the "bilinear identities". To introduce this identities we will show how they work when one considers quasi periodic solutions: this is the argument due to Cherednik [24]. Let us consider an algebraic curve Γ with genus g and select one point which we denote by ∞ and let k^1 be a local parameter (namely $k = \infty$ corresponds to the above selected point). Moreover let us choose $2g$ points $P_1, \ldots, P_g, P_1^*, \ldots, P_g^*$ so that $P_1 + \ldots + P_g + P_1^* + \ldots + P_g^* - 2\infty$ is linearly equivalent to the canonical divisor, namely a one-form ϑ exist such that:

$$(\vartheta) = P_1 + \cdots + P_g + P_1^* + \cdots + P_g^* - 2\infty \tag{5.3.1}$$

i.e. there is a one-form ϑ on Γ which has simple zeros at $P_1, \ldots, P_g^*$ and a double pole at infinity (in general, for a given set $P_1, \ldots, P_g$ one can find uniquely $P_1, \ldots, P_g$). Then we consider the following meromorphic functions $w(P)$ and $w^*(P)$ on $\Gamma - \{\infty\}$ normalized at infinity as follows:

$$\begin{aligned} w(x, P) &= \left(1 + \frac{w_1(x)}{k} + \cdots\right) e^{\xi(x,k)} \\ w^*(x, P) &= \left(1 + \frac{w_1^*(x)}{k} + \cdots\right) e^{-\xi(x,k)} \end{aligned} \tag{5.3.2}$$

Eqs. (5.3.2) have the same form required in the linearization of KP equation. The poles of $w(x, P)$ are in $P_1, \ldots, P_g$ and are simple poles while $w^*(x, P)$ is asked to have poles at $P_1^*, \ldots, P_g^*$. By these requirements $w(x, P)$ and $w^*(x, P)$ exist, and are uniquely determined. The following identity between $w(x, P)$ and $w^*(x, P)$ holds:

$$\int w(x, P) w^*(x', P) \vartheta = 0 \tag{5.3.3}$$

(5.3.3) can be derived by means of the residue theorem remembering that the zeros of ϑ cancel the poles of $w(x, P)$ and $w^*(x, P)$. Eq. (5.3.3) is the bilinear identity for the case of the so called "quasi periodic solution". We will show now that such a bilinear identity leads to the linear equations of KP type. One can find a differential operator $B_n(x, \partial_1)$ of order n in x_1 such that:

$$[\partial_n - B_n(x, \partial_1)] w(x, P) = O\left(\frac{1}{k}\right) e^{\xi(x,k)} \tag{5.3.4}$$

Such an operator exists due to the form (5.3.2) of $w(x, P)$, once we note that the action of ∂_1^n and ∂_n have the same effect on $e^{\xi(x,k)}$ in that both give a k^n factor: this makes it possible to subtract suitably the effect of the derivatives, leading to eq. (5.3.4) for a $B_n(x, \partial_1)$. If one applies the operator in square brackets in eq. (5.3.4) to $w(x, P)$ in the bilinear identity (5.3.3) and then differentiates $w^*(x', P)$ with respect to x'_1 and puts $x = x'$, one obtains:

$$\int [\partial_n - B_n(x, \partial_1)] w(x, P) \cdot \partial_1^j w^*(x, P) \vartheta = 0 \qquad j = 0, 1, 2, \ldots \quad (5.3.5)$$

In eq. (5.3.5) the exponential factors have disappeared in the product of w and w^*: the first factor is so $O(\frac{1}{k})$ by eq. (5.3.4) while, for $j = 0$, the rest has a double pole at infinity. Thus the integrand has a simple pole, but the residue must be zero and so the r.h.s. of eq. (5.3.4) should be in fact $O(\frac{1}{k^2}) e^{\xi(x,k)}$. Applying eq. (5.3.5) iteratively for $j = 1, 2$ etc. we see that all the residues for $1/k^{j+1}$ must be zero, obtaining:

$$[\partial_n - B_n(x, \partial_1)] w(x, P) = 0 \quad (5.3.6)$$

In this way we have shown that the bilinear identity leads to the linear KP hierarchy equations.

We will show now the equivalence between bilinear equation and bilinear identity. We need the following notation:

$$e^{\xi(x,k)} = \sum_{n=0}^{\infty} k^n p_n(x) \quad (5.3.7)$$

where $p_n(x)$ are polynomials:

$$\begin{aligned} p_0(x) &= 1 \\ p_1(x) &= x_1 \\ p_2(x) &= x_2 + \frac{x_1^2}{2} \\ p_3(x) &= x_3 + x_2 x_1 + \frac{x_1^3}{3} \end{aligned} \quad (5.3.8)$$

By means of $p_n(x)$ polynomials we can actually write all the Hirota differential equations for the KP hierarchy:

$$\sum_{j=0}^{\infty} p_j(-2y) p_{j+1}(\tilde{D}_x) \exp\Big\{\sum_{l=1}^{\infty} y_l D_{x_l}\Big\} \tau(x) \cdot \tau(x) = 0 \quad (5.3.9)$$

The hierarchy of KP equations can be obtained from eq. (5.3.9) equating to zero every coefficient of monomials of the arbitrary variable y. Now let us introduce the new vertex operators:

$$X(k) = e^{\xi(x,k)}e^{-\xi(\tilde{\partial},k^{-1})}$$
$$X^*(k) = e^{-\xi(x,k)}e^{\xi(\tilde{\partial},k^{-1})} \tag{5.3.10}$$

The two methods, i.e. linearization and bilinearization are connected by the following relations between w and τ functions:

$$w(x,k) = \tau^{-1}(x)X(k)\tau(x)$$
$$w^*(x,k) = \tau^{-1}(x)X^*(k)\tau(x) \tag{5.3.11}$$

which can be written in a more explicit form as follows:

$$w(x,k) = \tau^{-1}(x)\tau\left(x_1 - \frac{1}{k}, x_2 - \frac{1}{2k^2}, x_3 - \frac{1}{3k^3}, \dots\right)e^{\xi(x,k)}$$
$$w^*(x,k) = \tau^{-1}(x)\tau\left(x_1 + \frac{1}{k}, x_2 + \frac{1}{2k^2}, x_3 + \frac{1}{3k^3}, \dots\right)e^{-\xi(x,k)} \tag{5.3.12}$$

The bilinear identity in terms of τ becomes:

$$\int_{k=\infty} \tau\left(x_1 - \frac{1}{k}, \dots\right)\tau\left(x_1' + \frac{1}{k}, \dots\right)e^{\xi(x-x',k)}dk = 0 \tag{5.3.13}$$

(Now we are considering a more general scheme where the parameter k is considered only locally near infinity: this contains as a special case that of a curve of genus g for quasi periodic solutions). Eq. (5.3.13) is nothing but the Hirota equation. If one chooses $x = x'$, the exponential in (5.3.13) vanishes: taking the residue of $1/k$, which must be equal to zero, one obtains the equation (although this is a trivial equation):

$$D_1\tau(x) \cdot \tau(x) = 0 \tag{5.3.14}$$

In order to calculate the $1/k$ term one must differentiate the first τ function in (5.3.13) with minus sign and the second with plus sign so getting Hirota's derivation rule. If one first differentiates with respect to x' several times and then fixes $x = x'$, then one obtains more complicated and non-trivial equations of the form of Hirota. Thus both linear and bilinear methods are equivalent to the bilinear identity (5.3.13), which then contains the whole theory. In this sense this identity can be understood as characterization of the KP hierarchy. Now we will show another bilinear identity which characterizes the group orbit in the representation of infinite dimensional Lie algebras. Let us first introduce free fermions which are indexed by integers, more precisely charged fermions satisfying the usual anticommutation relations:

$$
\begin{aligned}
&[\psi_m, \psi_n]_+ = [\psi^*_m, \psi^*_n]_+ = 0 \\
&[\psi_m, \psi^*_n]_+ = \delta_{m,n} \qquad\qquad (m, n \in \mathbf{Z})
\end{aligned} \tag{5.3.15}
$$

and consider the Lie algebra of the following quadratic elements:

$$
[\psi_m \psi^*_n, \psi_{m'} \psi^*_{n'}] = \delta_{nm'} \psi_m \psi^*_{n'} - \delta_{mn'} \psi_{m'} \psi^*_n \tag{5.3.16}
$$

The rule (5.3.16) is nothing but the rule of the following kind of infinite dimensional matrices:

$$
\psi_m \psi^*_n \longleftrightarrow m \begin{pmatrix} & \overset{n}{\vdots} & \\ \cdots & 1 & \cdots \\ & \vdots & \end{pmatrix} \tag{5.3.17}
$$

(matrices in (5.3.17) have only one non zero entry equal to 1).

In this sense we call the algebra of relations (5.3.16) $\mathcal{GL}(\infty)$. One should be very careful. We have said that in the Lepowsky-Wilson work on vertex representation of $A_1^{(1)}$ the crucial point was that it contains a Heisenberg subalgebra. If we consider only finite sums of the above quadratic form, we cannot have a Heisenberg algebra because they are locally finite dimensional and we cannot expect a Heisenberg commutation relation. So we need infinite sums: but this can be very dangerous and one can very easily get contradictions. The simplest way to avoid such contradictions is to fix a representation. To this aim, we consider the Fock representation, namely we select a special vector $\mid 0>$ which we call "vacuum", or "highest weight vector" characterized (up to a constant multiple) by the following annihilation relations:

$$
\begin{aligned}
\psi_n \mid 0> &= 0 \qquad n < 0 \\
\psi^*_n \mid 0> &= 0 \qquad n \geq 0
\end{aligned} \tag{5.3.18}
$$

We introduce also a dual vacuum satisfying the complementary relations:

$$
\begin{aligned}
<0 \mid \psi_n &= 0 \qquad n \geq 0 \\
<0 \mid \psi^*_n &= 0 \qquad n < 0
\end{aligned} \tag{5.3.19}
$$

We normalize then as follows:

$$
<0 \mid 0> = 1 \tag{5.3.20}
$$

We introduce also the normal product:

$$
:\psi_m \psi^*_n: = \psi_m \psi^*_n - <0 \mid \psi_m \psi^*_n \mid 0> \tag{5.3.21}
$$

Now we consider only infinite sums of the form:

$$\chi = \sum_{m,n} c_{mn} : \psi_m^* \psi_n : \tag{5.3.22}$$

where $\exists N$ s.t. $c_{mn} = 0$ if $|m-n| > N$. This kind of operators act on the Fock space $\mathcal{F}$ without producing any divergence.

We introduce a "charge operator":

$$H_0 = \sum_{n \in \mathbf{Z}} : \psi_n \psi_n^* : \tag{5.3.23}$$

The Fock space can be divided into "charge sector" subspaces $\mathcal{F}_l$:

$$\mathcal{F}_l = \{ v \in \mathcal{F} \ |, H_0 v = l v \} \tag{5.3.24}$$

l is nothing but the number of ψ operators minus that of ψ^* acting on $|\,0>$ to obtain v.

We now introduce the "generalized vacua":

$$|\,l> = \begin{cases} \psi_{l-1} \cdots \psi_0 \,|\,0> & l > 0 \\ |\,0> & l = 0 \\ \psi_l^* \cdots \psi_{-1}^* \,|\,0> & l < 0 \end{cases} \tag{5.3.25}$$

with the normalization conditions:

$$<l\,|\,l'> = \delta_{ll'} \tag{5.3.26}$$

There is an automorphism which shifts charges:

$$\begin{aligned} \iota(\psi_m) &= \psi_{m+1} \\ \iota(\psi_m^*) &= \psi_{m+1}^* \\ \iota(\,|\,l>\,) &= |\,l+1> \end{aligned} \tag{5.3.27}$$

The following commutative dyagram holds:

$$\begin{array}{ccc} \mathcal{F} & \xrightarrow{\ \iota\ } & \mathcal{F} \\ {\scriptstyle a}\big\downarrow & & \big\downarrow{\scriptstyle \iota(a)} \\ \mathcal{F} & \xrightarrow[\ \iota\ ]{} & \mathcal{F} \end{array} \tag{5.3.28}$$

where a is any element of the fermion algebra.

In this sense nothing is different with different values of l.

The bilinear identity in this scheme can be written as follows. For an element v in $\mathcal{F}$:

$$\sum_{n\in\mathbf{Z}} \psi_n v \otimes \psi_n^* v = 0 \tag{5.3.29}$$

if and only if

$$v \in \bigcup_l \mathcal{GL}(\infty) \mid l > \tag{5.3.30}$$

The bilinear identity (5.3.29) characterizes the group orbit of generalized vacua. We briefly sketch the proof of (5.3.29). First let us consider only $v = \mid 0 >$-vector (we have already said that this is general enough due to the existence of ι automorphism). With this choice (5.3.29) is trivially true because either ψ_n or ψ_n^* annihilate $\mid 0 >$ due to eqs. (5.3.18):

$$\sum_{n\in\mathbf{Z}} \psi_n \mid 0 > \otimes \psi_n^* \mid 0 >= 0 \tag{5.3.31}$$

Now let us apply a Lie algebra element $X \in \mathcal{GL}(\infty)$ to the identity (5.3.31) using the commutation relations:

$$[X, \psi_n] = \sum_m \psi_m a_{mn} \tag{5.3.32}$$

$$[X, \psi_n^*] = -\sum_m \psi_m^* a_{nm} \tag{5.3.33}$$

We obtain the identity:

$$\sum_{n\in\mathbf{Z}} [X, \psi_n] \mid 0 > \otimes \psi_n^* \mid 0 > + \sum_{n\in\mathbf{Z}} \psi_n \mid 0 > \otimes [X, \psi_m^*] \mid 0 >= 0 \tag{5.3.34}$$

Eq. (5.3.34) says that the original identity (5.3.31) does not change by applying exponential of X: so for any vector which is on the group orbit the bilinear eq. (5.3.29) is satisfied, i.e. we have shown that (5.3.29) holds if v belongs to the group orbit. To prove that the bilinear identity holds only if v belongs to the group orbit, suppose $v = \sum_m v_m$ ($H_0 v_m = m v_m$) with $v_m \neq 0$, $v'_m \neq 0$, $m \neq m'$: we will show that this leads to contradictions. In general we can assume that $m = 0$ ($v_0 \in \mathcal{F}_0$). The bilinear identity does not change by the group action: so we can shift v_m by the group to an element of the form $v_m = \mid 0 > + \psi\psi\psi^*\psi^* \mid 0 > + \cdots$ (the quadratic part can be dropped by the group action).

From (5.3.31) one has to satisfy the two relations:

$$\sum_n \psi_n v_m \otimes \psi_n^* v_{m'} = 0 \tag{5.3.35}$$

$$\sum_n \psi_n v_{m'} \otimes \psi_n^* v_m = 0 \tag{5.3.36}$$

In eq. (5.3.35) $\psi_n v_m \neq 0$, $n \geq 0$ so $\psi_n^* v_{m'} = 0$ and in eq. (5.3.36) $\psi_n^* v_m \neq 0$, $n < 0$ so $\psi_n v_{m'} = 0$: the relations obtained for $v_{m'}$ show that it must be a constant multiple of the vacuum. This argument shows, in fact, that v belongs not only to some $\mathcal{F}_0$ but also to $\mathcal{GL}(\infty) \mid 0 >$. So in general if one asks a bilinear identity of the form (5.3.29) v should belong to one of the group orbits of $\mathcal{GL}(\infty)$.

The final task is now to establish the relation between the bilinear identities (5.3.29) and (5.3.13). Let us introduce the generating functions of free fermions:

$$\psi(k) = \sum_{n \in \mathbf{Z}} k^n \psi_n \tag{5.3.37}$$

$$\psi^*(k) = \sum_{n \in \mathbf{Z}} k^{-n} \psi_n^* \tag{5.3.38}$$

By means of (5.3.37) and (5.3.38) the bilinear identity (5.3.29) can be written in the integral form:

$$\int_{k=\infty} \frac{dk}{2\pi i k} \psi(k) g \mid 0 > \otimes \psi^*(k) g \mid 0 >= 0 \tag{5.3.39}$$

Now because of the tensor product, we can apply any vector $< a \mid$ and $< b \mid$ to eq. (5.3.39) as follows:

$$\int_{k=\infty} \frac{dk}{2\pi i k} < a \mid \psi(k) g \mid 0 > \otimes < b \mid \psi^*(k) g \mid 0 >= 0 \tag{5.3.40}$$

(We obviously choose $< a \mid \in \mathcal{F}_1^*$ and $< b \mid \in \mathcal{F}_{-1}^*$). The identity (5.3.40) is very similar to (5.3.13): the x and x' dependence should be contained in $< a \mid$ and $< b \mid$ in a suitable manner. We will show how to do this in the following section.

5.4 Fock and vertex representations

In this section we will show how to connect the two bilinear identities:

$$\int_{k=\infty} dk\, w(x,k) w^*(x',k) = 0 \tag{5.4.1}$$

$$\int \frac{dk}{2\pi i k} \psi(k) g \mid 0 > \otimes\, \psi^*(k) g \mid 0 > = 0 \tag{5.4.2}$$

where in the first identity the w-function is related to the τ-function by the relation:

$$w(x,k) = \tau^{-1}(x)\tau(x_1 - \frac{1}{k}, x_2 - \frac{1}{2k^2}, \dots) e^{\xi(x,k)} \tag{5.4.3}$$

Therefore our conclusion will be that the KP hierarchy, characterized by the bilinear identity (5.4.1), will be identified with the group orbit in the Fock representation whose equation is given by (5.4.2).

In the Fock space the charge l sector is given by:

$$\mathcal{F}_l = \oplus_{l=s-r} \mathbb{C}\ \psi^*_{m_1} \cdots \psi^*_{m_r} \psi_{n_s} \cdots \psi_{n_1} \mid 0 > \tag{5.4.4}$$

$$m_1 < \cdots < m_r < 0 \leq n_s < \cdots < n_1$$

(In eq. (5.4.4) s is the number of ψ fermions and r is that of ψ^* and their difference is l: the ordering of indices comes from the action of the fermions on the vacuum and from the antisymmetry of the product). The first problem is to identify the linear vector space (5.4.4) with the polynomial algebra. In order to do that, the crucial point is that $\mathcal{GL}(\infty)$ – the Lie algebra of bilinear free fermion products – contains the Heisenberg subalgebra $\mathcal{H}$. The H_j spanning $\mathcal{H}$ has the fermion realization:

$$H_j = \sum_{n \in \mathbf{Z}} \psi_n \psi^*_{n+j} \qquad (j \neq 0) \tag{5.4.5}$$

They satisfy the same commutation relations of the differential operators:

$$H_j = \begin{cases} \frac{\partial}{\partial x_j} & j > 0 \\ -j x_{-j} & j < 0 \end{cases} \tag{5.4.6}$$

Now we want to find a linear map between $\mathcal{F}_l$ and the polynomial algebra $\mathbb{C}\left[x_1, x_2, x_3, \dots\right]$ which is the natural representation space of the Heisenberg algebra. It is given by taking the inner product with an element $< v^*_l \mid \in \mathcal{F}^*_l$ satisfying:

$$\begin{aligned} \partial_j < v_l^* \mid &= < v_l^* \mid H_j \qquad (j > 0) \\ -jx_{-j} < v_l^* \mid &= < v_l^* \mid H_j \qquad (j < 0) \end{aligned} \tag{5.4.7}$$

The solution of eq. (5.4.7) is unique and is given by the vector (which contains, of course, the x dependence):

$$\begin{aligned} < v_l^* \mid &= < l \mid e^{H(x)} \\ H(x) &= \sum_{j=1}^{\infty} x_j H_j \end{aligned} \tag{5.4.8}$$

where:

$$< l \mid = \begin{cases} < 0 \mid \psi_0^* \ldots \psi_{l-1}^* & l > 0 \\ < 0 \mid & l = 0 \\ < 0 \mid \psi_{-1} \ldots \psi_{-l} & l < 0 \end{cases} \tag{5.4.9}$$

So to any vector in the Fock space $\mathcal{F}_l$ we associate a function of the variables x:

$$a \mid l > \longrightarrow < l \mid e^{H(x)} a \mid l > \tag{5.4.10}$$

This correspondence between the charge l sector of Fock space $\mathcal{F}_l$ and the polynomial algebra $\mathbb{C}\left[x_1, x_2, \ldots\right]$ is nothing but the usual "fermion-boson correspondence". This correspondence is actually an isomorphism. The proof is very simple. First of all it is "onto" because the Heisenberg algebra acts already irreducibly. If one counts the degree one finds that it is also injective. Still it is desiderable to obtain a more explicit correspondence between these two spaces. In order to do that we introduce and compute the Schur polynomial, i.e. the character polynomial.

To this end we recall some basic results from representation theory of finite dimensional Lie algebras. Let us consider the highest weight representation of sl$(n, \mathbb{C})$ (N is taken to be a sufficiently large number). sl$(N, \mathbb{C})$ can be decomposed into three parts:

$$\mathrm{sl}(N, \mathbb{C}) = n_+ \oplus \mathrm{h} \oplus n_- \tag{5.4.11}$$

where the first and third parts correspond to upper and lower triangular matrices respectively, while the second part h corresponds to diagonal matrices (it is called Cartan subalgebra).

The (irreducible) highest weight representations are constructed by means of a vacuum vector v_Λ which is annihilated by n_+:

$$n_+ v_\Lambda = 0 \tag{5.4.12}$$

v_Λ generates all the other vectors by means of the action of n_-, and is an eigenvector of the Cartan subalgebra:

$$hv_\Lambda = \Lambda(h)v_\Lambda \qquad \forall h \in \mathrm{h}, \Lambda \in \mathrm{h}^* \qquad (5.4.13)$$

where we denote by h^* the dual Cartan subalgebra. We require that Λ is integral dominant, namely if h_j is a basis for h:

$$h_j = \begin{pmatrix} & & j & \\ & \vdots & & \\ & 1 & & \\ & & & -1 \end{pmatrix} \qquad (5.4.14)$$

and Λ_j is the cobasis corresponding to h_j:

$$\Lambda_j(h_i) = \delta_{ji} \qquad (5.4.15)$$

then Λ is given by:

$$\Lambda = \sum l_j \Lambda_j \qquad (5.4.16)$$

where l_j are non negative integers.

For any such $\Lambda \in \mathrm{h}^*$ there is a unique irreducible representation which satisfies conditions (5.4.12) and (5.4.13).

To each highest weight representation there corresponds a Young diagram which appears like in fig. 5.4.1

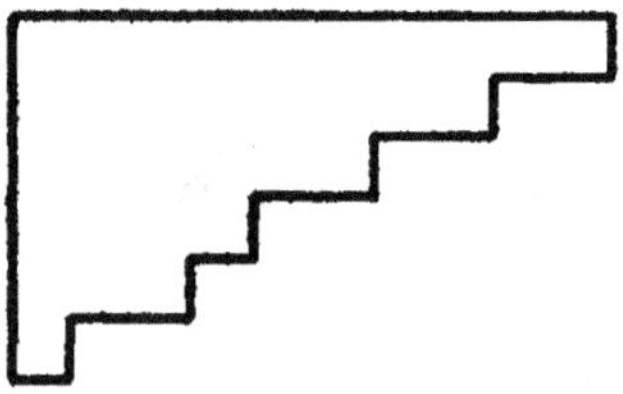

fig. 5.4.1

We construct the Young dyagram Y by drawing, in decreasing order, rectangles whose lenght is j, namely the index of the weight Λ_j, and we draw for any j so many rectangles as the positive integer l_j. For example:

$\Lambda_1 \longleftrightarrow \square$

$2\Lambda_1 \longleftrightarrow \square\square$

$\Lambda_2 \longleftrightarrow \begin{array}{c}\square\\\square\end{array}$

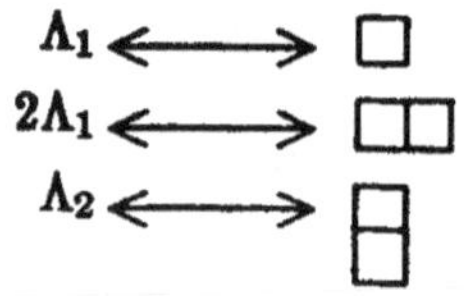

fig. 5.4.2

Now the Schur polynomial is nothing but the usual character, i.e. the trace of the group element in the considered representation. It is sufficient to consider only the Cartan part of the group. The Schur polynomial uses the following independent variables:

$$x_j = \frac{1}{j} \sum_{m=1}^{N} e^{j\epsilon_m}, \quad h = \begin{pmatrix} \epsilon_1 & & \\ & \ddots & \\ & & \epsilon_N \end{pmatrix} \in \mathrm{h} \tag{5.4.17}$$

So for the representation ρ_Y corresponding to the Young dyagram Y, the Schur polynomial χ_Y is:

$$\mathrm{trace}\, \rho_Y(e^h) = \chi_Y(x) \tag{5.4.18}$$

where $x = (x_1, x_2, \ldots)$ and h are given by eq. (5.4.17).

The Schur polynomial can be computed by means of Weyl's formula:

$$e^{-\xi(x,k)} = \sum_n q_n(x)(-k)^n \tag{5.4.20a}$$

$$\xi(x,k) = \sum x_n k^n \tag{5.4.20b}$$

$$\chi_Y(x) = \det \begin{bmatrix} q_{k_1}(x) q_{k_1+1}(x) & & \cdots \\ & \ddots & \\ \cdots & & q_{k_l-1}(x) q_{k_l}(x) \end{bmatrix} \tag{5.4.20c}$$

$Y =$ 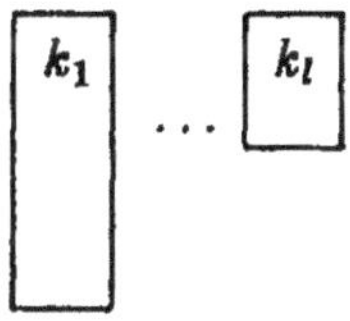

The form of the polynomials $\chi_Y(x)$ does not depend on N, for N sufficiently large. So $\chi_Y(x)$ are well defined polynomials of the form:

$$\begin{aligned} \chi_{\square}(x) &= x_1 \\ \chi_{\square\square}(x) &= x_2 + \frac{1}{2} x_1^2 \\ \chi_{\genfrac{}{}{0pt}{}{\square}{\square}}(x) &= -x_2 + \frac{1}{2} x_1^2 \end{aligned} \tag{5.4.21}$$

Note that $\xi(x,k)$ is just the expansion introduced in the previous lectures and that the polynomials $q_j(x)$ are simply connected to the polynomials $p_j(x)$ defined in eq. (5.3.7):

$$p_j(x) = (-)^j q_j(-x) \tag{5.4.22}$$

Now we will rewrite Weyl's formula in terms of free fermions: this will give us the correspondence between the Fock space and the polynomial algebra. The Schur polynomials constitute a basis of the polynomial ring. So the Weyl formula in terms of free fermions will be useful in identifying the polynomial algebra with the Fock space. The most general formula can be written in the following way. For given r and s let us consider the Young diagram of fig. 5.4.3. The character polynomial is then given by the free fermions expectation values:

$$\chi_Y = (-)^{m_1+\cdots+m_r+(s-r)(s-r-1)/2}\times$$
$$\times < s-r \mid e^{H(x)}\psi^*_{m_1}\cdots\psi^*_{m_r}\psi_{n_s}\cdots\psi_{n_1} \mid 0 > \tag{5.4.23}$$

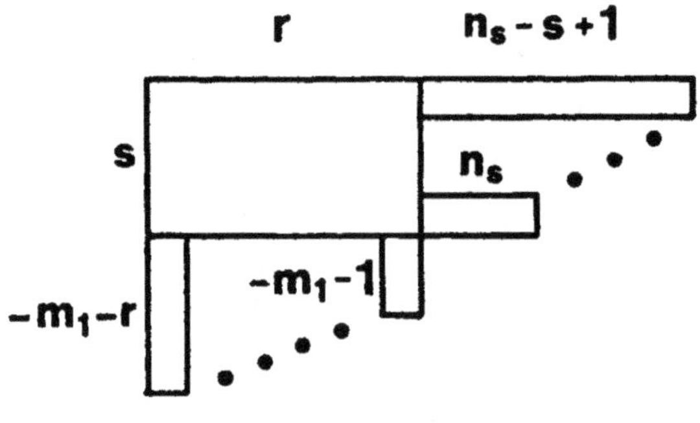

fig. 5.4.3

Eq. (5.4.23) is nothing but the formula (5.4.10). We give now a sketch of the proof of eq. (5.4.23). Let us consider the generating functions of free fermions and their "time evolution" by the "Hamiltonian" $H(x)$:

$$e^{H(x)}\psi(k)e^{-H(x)} = e^{\xi(x,k)}\psi(k) \tag{5.4.24a}$$
$$e^{H(x)}\psi^*(k)e^{-H(x)} = e^{-\xi(x,k)}\psi^*(k) \tag{5.4.24b}$$

(The "time evolution" is diagonalized by the generating functions $\psi(k)$ and $\psi^*(k)$). The Hamiltonian annihilates the vacuum:

$$H(x) \mid 0 >= 0 \tag{5.4.25}$$

Taking this last relation in mind we let the operators in eq. (5.4.24 b) act on the vacuum, expand the generating function $\psi^*(k)$ on both sides and make a scalar product with $< 0 \mid \psi_j$. The r.h. side of (5.4.24 b) gives us just the coefficients of k^l in eq. (5.4.20 a), so we obtain:

$$(-)^l q_l(x) = < 0 \mid \psi_j e^{H(x)} \psi^*_{j-l} \mid 0 > \qquad (5.4.26)$$

Eq. (5.4.26) can be rewritten by means of the Wick theorem in term of the following matrix element:

$$M_l(x) = < 0 \mid \psi_{-1} \cdots \psi_{-l} \, e^{H(x)} \psi^*_{m_1} \cdots \psi^*_{m_l} \mid 0 > \qquad (5.4.27)$$

The matrix element in eq. (5.4.27) is just that of eq. (5.4.23) for $r = -l$, $s = 0$. The general case can be treated as follows: we will explain it by means of an example. Let us consider the case $s = r = 1$ and the Young diagram shown in fig 5.4.4:

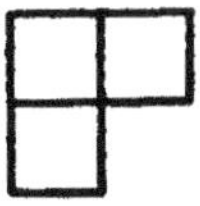

fig. 5.4.4

One should calculate the matrix element:

$$M = < 0 \mid e^{H(x)} \psi^*_{-2} \psi_1 \mid 0 > \qquad (5.4.28a)$$

We can shift charges in the matrix element (5.4.28 a) by -2:

$$M = < -2 \mid e^{H(x)} \psi^*_{-4} \psi_{-1} \mid -2 > \qquad (5.4.28b)$$

Using the definition of $\mid -2 >$ we obtain:

$$M = < -2 \mid e^{H(x)} \psi^*_{-4} \psi_{-1} \psi^*_{-2} \psi^*_{-1} \mid 0 > \qquad (5.4.28c)$$

In eq. (5.4.28 c) ψ_{-1} and ψ^*_{-1} cancel each other giving a minus sign:

$$M = (-) < -2 \mid e^{H(x)} \psi^*_{-4} \psi^*_{-2} \mid 0 > \qquad (5.4.28d)$$

The general case can be proved by this same shift argument (it is also easy to see that the matrix element corresponds to the same Young diagram of fig. 5.4.4 using eq. (5.4.23)).

We have so established not only the equivalence but also the explicit correspondence between the $\mathcal{F}_l$ charge sector Fock space and the polynomial algebra $\mathbf{C}\left[x_1, x_2, \ldots\right]$. So the action of H_j, by multiplying on the left, can be realized, in this language, by differentiation and multiplication:

$$H_j = \sum_{n \in \mathbf{Z}} \psi_n \psi^*_{n+j} \longrightarrow \begin{cases} \frac{\partial}{\partial x_{-j}} & j > 0 \\ -j x_{-j} & j < 0 \end{cases} \tag{5.4.29}$$

The natural question is now how the general element of the Lie algebra $\mathcal{GL}(\infty)$ is realized on this bosonic space: the answer can be given by means of the vertex operators. We recall the definition of the vertex operators:

$$\begin{aligned} X(k) &= e^{\xi(x,k)} e^{-\xi(\tilde{\partial}, k^{-1})} \\ X^*(k) &= e^{-\xi(x,k)} e^{\xi(\tilde{\partial}, k^{-1})} \end{aligned} \tag{5.4.30}$$

It is easy to see that there is a simple correspondence between $\psi(k)$ and the vertex operator $X(k)$ given by the following commutative diagrams:

$$\begin{array}{ccc} \mathcal{F}_{l-1} & \longrightarrow & \mathbf{C}\left[x_1, x_2, \ldots\right] \\ \Big\downarrow \psi(k) & & \Big\downarrow k^{l-1} X(k) \\ \mathcal{F}_l & \longrightarrow & \mathbf{C}\left[x_1, x_2, \ldots\right] \\ \Big\downarrow \psi^*(k) & & \Big\downarrow k^{-l+1} X^*(k) \\ \mathcal{F}_{l-1} & \longrightarrow & \mathbf{C}\left[x_1, x_2, \ldots\right] \end{array} \tag{5.4.31}$$

Let us briefly sketch the proof of the correspondence between $\psi(k)$ and the vertex operator $X(k)$ given in the diagrams (5.4.31).

We will consider the particular case:

$$\begin{array}{ccc} \mathcal{F}_1 & \longrightarrow & \mathbf{C}\left[x_1, x_2, \ldots\right] \\ \Big\downarrow \psi^*(k) & & X^*(k) \Big\downarrow \\ \mathcal{F}_0 & \longrightarrow & \mathbf{C}\left[x_1, x_2, \ldots\right] \end{array} \tag{5.4.32}$$

So we have to show the following identity:

$$\begin{aligned} &< 0 \mid e^{H(x)}\psi^*(k)\psi^*(p_1)\cdots\psi^*(p_s)\psi(q_s)\cdots\psi(q_0) \mid 0 >= \\ &= X^*(k) < 0 \mid \psi_0^* e^{H(x)}\psi^*(p_1)\cdots\psi^*(p_s)\psi(q_s)\cdots(q_0) \mid 0 > \end{aligned} \tag{5.4.33}$$

We can see that the left side of eq. (5.4.33) represents the action of $\psi^*(k)$ on charge 1 sector; the right side must be identifyed with the action of the vertex operator on an element of the polynomial algebra (remembering that $< 1 \mid = < 0 \mid \psi_0^*$). We will use the same kind of argument used in the previous section when we considered the quasi-periodic solutions. Both sides of (5.4.33) are meromorphic in k except at infinity.

The left side has zeros at $k = p_1, \ldots, p_s$ because there we would have the square of a free fermion; it has also poles at $k = q_s, \ldots, q_0$ because one has the expectation value:

$$< \psi^*(p)\psi(q) >= \frac{p}{p-q} \tag{5.4.34}$$

We have to show that the right side of eq. (5.4.33) has the same zeros and poles. If we shift the exponential $e^{H(x)}$ through fermions up to the vacuum we obtain a product of factors of the form $e^{-\xi(x,p_1)}e^{\xi(x,q_s)}$ and the exponential $e^{H(x)}$ vanishes as $H(x)$ annihilates the vacuum. The action of the vertex operator on the exponential factor thus obtained is given by:

$$\begin{aligned} e^{\mp\xi(\tilde{\partial},k^{-1})}e^{\pm\xi(x,p)} &= \left(1 - \frac{p}{k}\right)e^{\pm\xi(x,p)} \\ e^{\pm\xi(\tilde{\partial},k^{-1})}e^{\pm\xi(x,p)} &= \left(1 - \frac{p}{k}\right)^{-1}e^{\pm\xi(x,p)} \end{aligned} \tag{5.4.35}$$

Eq. (5.4.35) says that also the right side of eq. (5.4.33) has the same zeros and poles of the left: so they differ at most by a multiplicative constant, but this can be calculated at $k = \infty$ where both sides become:

$$
\begin{aligned}
(5.4.33) \xrightarrow[k=\infty]{} & \left(< 0 \mid \psi_0^* \psi^*(p_1) \cdots \psi^*(q_0) \mid 0 > + O(\frac{1}{k}) \right) \times \\
& \times \exp \left\{ -\xi(x,k) - \sum_{j=1}^{s} \xi(x,p_j) + \sum_{j=0}^{s} \xi(x,q_j) \right\}
\end{aligned}
\qquad (5.4.36)
$$

We have so identified every sector of the Fock space with the polynomial algebra $\mathbb{C}\left[x_1, x_2, \ldots\right]$ and we are able to compute the effect of the action of the generic Lie algebra $\mathcal{GL}(\infty)$ element by using the previous formulas:

$$
\begin{array}{ccc}
\mathcal{F}_0 & \longrightarrow & \mathbb{C}\left[x_1, x_2, \ldots\right] \\
\Big\downarrow \psi(p)\psi^*(q) & & \frac{q}{p-q} X(p,q) \Big\downarrow \\
\mathcal{F}_0 & \longrightarrow & \mathbb{C}\left[x_1, x_2, \ldots\right]
\end{array}
\qquad (5.4.37)
$$

This explains what was said in the first section, i.e. the N-soliton solution can be obtained by means of the vertex operator as follows:

$$
\exp \left\{ \sum_{j=1}^{N} a_j X(p_j, q_j) \right\} \cdot 1 = \tau_N \qquad (5.4.38)
$$

In fact, in this language the τ function is nothing but the expectation value:

$$
\tau(x) = < 0 \mid e^{H(x)} g \mid 0 > \qquad (5.4.39)
$$

(g is a group element; $g \in \mathcal{GL}(\infty)$), while the wave functions are given by:

$$
\begin{aligned}
w(x,k) &= < 1 \mid e^{H(x)} \psi(k) g \mid 0 > \\
w^*(x,k) &= < -1 \mid e^{H(x)} \psi^*(k) g \mid 0 >
\end{aligned}
\qquad (5.4.40)
$$

Eqs. (5.4.40) show the equivalence between the two bilinear identities (5.4.1 — 5.4.2) which now are summarized as follows:

$$\int \frac{dk}{2\pi i k} < 1 \mid e^{H(x)}\psi(k)g \mid 0 >< -1 \mid e^{H(x)}\psi^*(k)g \mid 0 >= 0 \quad (5.4.41)$$

So eq. (5.4.41), which characterizes the Fock representation of the group orbit of $\mathcal{GL}(\infty)$ gives also the characterization of the solution space of KP hierarchy.

Before ending we will show how the various types of soliton equations can be explained in this language, without giving details.

We can obtain several classes of soliton equations by means of l-reduction and of the Kac-Moody Lie algebra $A_{l-1}^{(1)} \subset \mathcal{GL}(\infty)$ which is generated by:

$$\begin{aligned} e_i &= \sum_{n \equiv i \bmod l} \psi_{n-1}\psi_n^* \\ f_i &= \sum_{n \equiv i \bmod l} \psi_{n-1}^* \end{aligned} \quad (5.4.42)$$

It is easy to see that the generators (5.4.42) belong to the kernel of H_{lj} (see definition (5.4.5)): this implies that the vertex operators corresponding to e_i and f_i do not contain x_{lj} or $\frac{\partial}{\partial x_{lj}}$. This makes the τ function indipendent of x_{lj}:

$$\frac{\partial \tau(x)}{\partial x_l} = \frac{\partial \tau(x)}{\partial x_{2l}} = \cdots = 0 \quad (5.4.43)$$

For the case $l = 2$ this is nothing but the condition for KdV, namely one obtains the KdV hierarchy dropping the dependence on even parameters in the KV hierarchy:

$$l = 2 \qquad \text{KdV} \qquad \frac{\partial u}{\partial x_3} = \frac{3}{2}u\frac{\partial u}{\partial x_1} + \frac{1}{4}\frac{\partial^3 u}{\partial x_1^3} \quad (5.4.44)$$

Similarly for $l = 3$, $A_2^{(1)}$ gives the Boussinesq equation:

$$l = 3 \quad \text{Bousinnesq} \quad 3\frac{\partial^2 u}{\partial x_2^2} + \frac{\partial}{\partial x_1}\left(6u\frac{\partial u}{\partial x_1} + \frac{\partial^3 u}{\partial x_1^3}\right) = 0 \quad (5.4.45)$$

Obviously $l \geq 2$, because for $l = 1$ one obtains trivially the Heisenberg algebra.

Then we can go further, considering subalgebras of $A_l^{(1)}$. In such a case it is more natural to introduce the following τ function:

$$\tau_n(x) = < n \mid e^{H(x)} g \mid n > \tag{5.4.46}$$

For $A_l^{(1)}$ the definition (5.4.46) does not give a new τ function because it contains only a different g and one can shift n to zero. We also note the invariance property for τ:

$$\tau_{n+l}(x) = \tau_n(x) \tag{5.4.47}$$

For a given subalgebra of $A_l^{(1)}$, different τ_n functions correspond to different types of soliton equations. From the point of view of representation theory of Lie algebras, different vacuum vectors in eq. (5.4.46) mean different highest weight representations, namely τ_n corresponds to the fundamental representation Λ_n. We will now consider some subalgebras of $A_l^{(1)}$. In particular we will consider involutions of the Lie algebras corresponding to the up-down involution σ of the Dynkin dyagram.

If we consider the invariant part with respect to this involution we obtain different types of Kac Moody Lie algebras (shown in fig. 5.4.5):

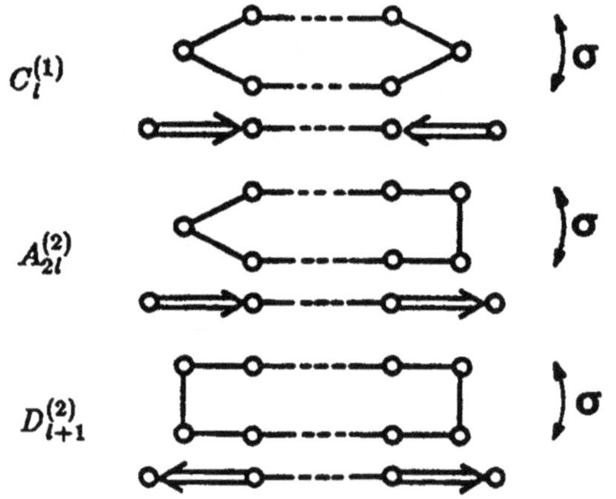

fig. 5.4.5

We want to see the effect of going to subalgebras in terms of soliton theory. The case $\mathcal{GL}(\infty) \to A_l^{(1)}$ gives the invariance properties (5.4.43) and (5.4.47). Similarly going further to the above subalgebras one obtains the invariance property:

$$\tau_n(x) = \tau_{\sigma(n)}(\tilde{x}) \tag{5.4.48}$$

where σ is the involution shown in fig. 5.4.6 and $\tilde{x}$ is obtained from x by negating all even parameters:

$$\tilde{x} = (x_1, -x_2, x_3, \ldots) \tag{5.4.49}$$

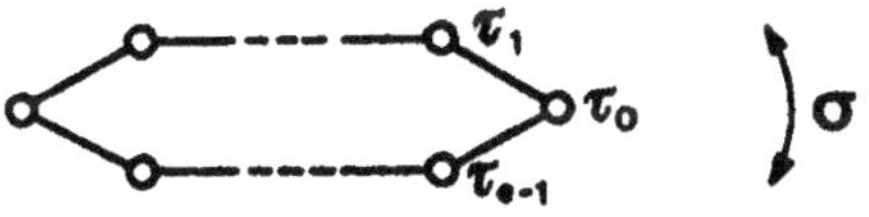

fig. 5.4.6

So different highest weight representations give different kinds of soliton equations (that for $A_2^{(2)}$ was already known to Sawada-Kotera and Kaup [25, 26]): this gives a systematic scheme of classification of soliton equations (we refer the reader also to the classification scheme by Drinfeld and Sokolov [27]).
To include other types of soliton equations we can do another generalization. Until now $k = \infty$ was a special point, i.e. it was an essential singularity: this gives a particular form for the "time evolution" (see for example eqs. (5.4.24) showing that $\psi(k)$ has a singularity at $k = \infty$). We can consider more general forms of the time evolution of eqs. (5.4.24), namely we can introduce singularities at finite points. We can then reconstruct the whole theory in a similar manner with this different time evolution. If we consider $\mathcal{GL}(\infty)$ with poles at $k = \infty$ and $k = 0$ we obtain the two dimensional Toda lattice:

$$\frac{\partial^2}{\partial x \partial y} u_m = -\sum_n a_{nm} e^{-u_n} \tag{5.4.50}$$

$$\{a_{nm}\} = \begin{bmatrix} \ddots & & & & & \\ \ddots & \cdot & & & & \\ \cdot & 2 & -1 & & & \\ & -1 & 2 & -1 & & \\ & & \cdot & \cdot & \cdot & \\ & & & \cdot & 2 & 1 \\ & & & & -1 & 2\cdot \\ & & & & & \ddots \end{bmatrix}$$

(a_{nm}: Cartan matrix of $\mathcal{GL}(\infty)$). If we restrict to $A_1^{(1)}$, the two dimensional Toda lattice reduces to Sine-Gordon equations: the only difference is the "time evolution". We can also consider a different time evolution (l, m, n discrete variables):

$$\begin{aligned} \psi(k) &\mapsto (1-ak)^{-l}(1-bk)^{-m}(1-ck)^{-n}e^{\xi(x_0,k)}\psi(k) \\ \psi^*(k) &\mapsto (1-ak)^{l}(1-bk)^{m}(1-ck)^{n}e^{-\xi(x_0,k)}\psi^*(k) \end{aligned} \tag{5.4.51}$$

(instead of $\psi(k) \mapsto k^n e^{\xi(x,k)+\xi(x,k^{-1})}\psi(k)$ for Toda lattice).

The time evolution with respect to the discrete variables l, m, n gives the discrete version of KP hierarchy (this was first found by Hirota: it is nothing but the KP hierarchy but in a difference equation form).

There is another generalization of the theory: we can consider also multicomponent fermions with n degrees of isospin:

$$\psi^{(i)}(k), \quad \psi^{(i)^*}(k) \qquad i = 1, \ldots, n \tag{5.4.52}$$

The natural condition for reduction is to equate the different sets of fermions. For the case $n = 2$ a suitable reduction condition gives again $A_1^{(1)}$ and the corresponding soliton equations are the non linear Schrödinger equations considering the time evolution at $k = \infty$, the chiral fields for $k = \pm 1$ and the Lund-Regge equation for $k = 0, \infty$.

We can consider also a completely different type of free fermions, e.g. neutral fermions. Let us introduce these very strange fermions. We consider an elliptic curve C defined by:

$$C: \quad \omega^2 = (k^2 - a^2)(k^2 - b^2) \tag{5.4.53}$$

$$P = (\omega, k) \in C$$

a and b are constants (in the former case of KP, the curve is just $\mathbf{P}^1$ and k is the local parameter). Similarly we introduce free fermions indexed by the point on the elliptic curve and a vacuum giving the following expectation value:

$$< \phi(P)\phi(P') > = \frac{1}{2}\frac{\omega + k^2 - \omega' - k'^2}{k + k'} \tag{5.4.54}$$

If we consider the two components reduced theory of these free fermions we obtain the Landau-Lifshitz equation:

$$i\frac{\partial \mathbf{S}}{\partial t} = \mathbf{S}_x \mathbf{S}_{xx} + \mathbf{J}\mathbf{S}_x \cdot \mathbf{S} \tag{5.4.55a}$$

$$\mathbf{S} = (S^1, S^2, S^3) \tag{5.4.55b}$$

$$\sum_{i=1}^{3} (S^i)^2 = 1\,, \qquad \mathbf{J} = \begin{pmatrix} J_1 & & \\ & J_2 & \\ & & J_3 \end{pmatrix}$$

The corresponding Lie algebra is $\mathrm{sl}(2, \mathbb{C}\,[k, k^{-1}, \omega]) \times \mathbb{C}_1$ In this way we can understand most soliton equations in terms of representation theory of Lie algebras.

References

[1] M. Jimbo, T. Miwa and M. Sato, Kakuyugo Kenkyu, Inst. Plasma Physics Nagoya Univ. **40**, Suppl., 45 (1978). E. Date, M. Kashiwara, M. Jimbo and T. Miwa, Proc. RIMS Symposium on "Non-Linear Integrable Systems: Classical Theory, Quantum Theory"; M. Jimbo and T. Miwa eds., World Scientific Publ. Co., Singapore (1983); p. 39.

[2] E. Barouch, B.M. McCoy and T.T.Wu, Phys. Rev. Lett. **31**, 1409 (1973)

[3] C.A. Tracy and B.M. McCoy, Phys. Rev. Lett. **31**, 1500 (1973)

[4] T.T. Wu, B.M. McCoy, C.A. Tracy and E. Barouch, Phys. Rev. **B13**, 316 (1976).

[5] L. Schlesinger, J. Reine u. Angew. Math. **141**, 96 (1912).

[6] L. Onsager, Phys. Rev. **65**, 117 (1944).

[7] B.M. McCoy, J.H.H. Perk and T.T. Wu, Phys. Rev. Lett. **46**, 757 (1981).

[8] E.H. Lieb and W. Liniger, Phys. Rev. **130**, 1605 (1963).

[9] L.A. Takhtadzhan and L.D. Fadeev, Russian Math. Surveys (Uspekhi Mat. Nouk)**34 5**, 11 (1979).

[10] H.B. Thacker, Rev. Mod. Phys. **53**, 253 (1981).

[11] J. Honerkamp, P. Weber and A. Wiesler, Nucl. Phys. **B152**, 266 (1979).

[12] D.B. Creamer, H.B. Thacker and D. Wilkinson, Gelfand-Levitan method for operator fields, FERMILAB preprint Pub. 79/75 THY (1979).

[13] H.B. Thacker and D. Wilkinson, Phys. Rev. **D19**, 3660 (1979).

[14] H.G. Vaidya and C.A. Tracy, Phys. Rev. Lett. **42**, 3 (1979).

[15] A. Lenard, J. Math. Phys. **7**, 1268 (1966).

[16] M. Kashiwara and T. Miwa, Proc. Japan Acad. **57A**, 342 (1981).

[17] E. Date, M. Kashiwara and T. Miwa, ibid. **57A**, 387 (1981).

[18] E. Date, M. Jimbo, M. Kashiwara and T. Miwa, J. Phys. Soc. Jpn **50** 3806, 3813 (1981).

[19] E. Date, M. Jimbo, M. Kashiwara and T. Miwa, Physica **4D**, 343 (1982); Publ. RIMS, Kyoto Univ. **18** 1077, 1111 (1982).

[20] M. Sato, Soliton equations as dynamical system on infinite dimensional Grassmann manifolds, RIMS Kokyuroku **439**, 30 (1981).

[21] M. Sato and Y. Sato (Mori), On Hirota's bilinear equations I, II, RIMS Kokyuroku **388**, 183 (1980), **414**, 181 (1981).

[22] I.M. Krichever, Russ. Math. Surveys **32**, 185 (1977).

[23] J. Lepowski and R.L. Wilson, Commun. Math. Phys. **62**, 43

(1978).

[24] I.V. Cherednik, Funct. Anal. and Its Appl. **12**, 195 (1978).

[25] K. Sawada and T. Kotera, Prog. Theo. Phys. **51**, 1355 (1974).

[26] J. Koup, Stud. in Appl. Math. **62**, 189 (1980). [27] V.G. Drinfeld and V.V. Sokolov, Dokl. Akad. Nouk USSR **258**, 11 (1980).

Potts models and dichromatic polynomials

PAUL P. MARTIN

Department of Physics
Queen Mary College
Mile End Road
London E1 4NS, England

1. Introduction

Most exact work in Statistical Mechanics concerns Ising-like models (i.e. those having Z_2 valued spin variables). It is important from a physical point of view, however, that a wider class of models eventually be addressed. To this end we will discuss the exact information presently available for the Q-state Potts model. In particular we will give evidence for a part circle distribution of zeros of the partition function in the complex exponentiated coupling constant plane close to the phase transition point.

It will be helpful to start with a brief review of existing work (an extended survey may be found, for instance, in [1]). The partition function for the Q-state Potts model may be written:

$$Z = \sum_{\substack{configurations \\ \{\sigma\}}} \exp\left(k \sum_{<ij>} \delta(\sigma_i, \sigma_j) \right) \tag{1}$$

where:

$$\delta(\sigma_i, \sigma_j) = \begin{cases} 1 & \text{if } \sigma_i = \sigma_j \\ 0 & \text{otherwise} \end{cases}$$

and σ_i takes Q distinct possible values on each vertex i of a lattice (the first sum is over possible configurations of these vertex spins on the lattice and the second sum is over edges on the lattice). Putting $v = e^k - 1$ then:

$$Z = \sum_{\{\sigma\}} \prod_{<ij>} \left(1 + v\delta(\sigma_i, \sigma_j)\right) = \sum_G Q^c v^l \equiv Z(Q, k) \tag{2}$$

where the sum over graphs G represents a sum over all possible choices of 1 or v on the edges $< ij >$. Because of the delta functions all sites connected by v's in a given graph must have the same σ, so c is the number of distinct clusters of such connected sites and l is the number of v's in a graph. This dichromatic polynomial is difficult to obtain for any given lattice because c is non-local. However it may be rewritten with local weights as a vertex model. Specifically, it can be shown [1, 2] that the Potts model partition function is equivalent to an ice-type model with different vertex weights on alternate sites of a medial lattice (constructed from the original lattice as shown in figure 1). The vertex weights may be written:

Vertex	1	2	3	4	5	6

type 1 weight(ω_i)	1	1	x_1	x_1	$1+Lx_1$	$1+x_1/L$
type 2 weight($\overline{\omega}_{\bar{i}}$)	x_2	x_2	1	1	x_2+L	x_2+1/L

where L is given by $\sqrt{Q} = (L+1/L)$ and $x = (1/\sqrt{Q})(e^k - 1)$. The type 1 weights are used at *odd* sites (corresponding to horizontal bonds of the original lattice, say) and type 2 at *even* sites (vertical bonds).

The partition function then becomes:

$$Z_{ice} = \sum_{\substack{configs.\\ \{\omega\}}} \Bigg(\prod_{\substack{odd\\ sites\\ i}} \omega_i \overline{\omega}_{\bar{i}} \Bigg) \Bigg(\prod_{\substack{external\\ sites}} L^{p/4} \Bigg) \tag{3}$$

($\bar{i}$ refers to an even site paired with the odd site i). The sum over configurations is a sum over possible arrow coverings of the bonds of the medial lattice and:

$$p = \begin{cases} 1 & \text{for clockwise arrows} \\ -1 & \text{otherwise.} \end{cases}$$

With all weights of type 1 such a model can be solved [1]. In the present case it cannot. However, putting $x_2 = 1/x_1$ and multiplying all type 2 weights by x_1 we recover the solvable model. Now, $x_2 = 1/x_1$ also gives the fixed point of the Potts model duality transformation for Q integer [3, 4] so at the dual point the Potts model reduces to a solved homogeneous model. This model turns out to be in its critical sector [1]. Furthermore, all eigenvalues of the transfer matrix for the Q-state model may be related to those for the mixed weight model [5]. We thus anticipate that the Potts model be critical for $x_2 = 1/x_1$. In homogeneous Potts model notation:

$$(e^{k_c}-1)^2 = Q \tag{4}$$

Similar arguments reveal one other set of critical values:

$$e^{k_c} = -1 \pm \sqrt{4-Q} \tag{5}$$

(see [6]).

2. *Global analytic structure of the partition function*

Existing methods of solution for the Ising model make such specific use of its symmetries that they cannot be generalised to other Potts models. However, there are some properties of the partition function which may be generalised. Now a convenient way to exhibit properties of a partition function is through the distribution of its roots in the complex exponentiated coupling constant plane. We will see that this procedure manifests similarities between different Potts model partition functions – in their global analytic structure.

We will illustrate what is meant by "global analytic structure" by the example of the Ising model. In this case [7]:

$$\lim_{N\to\infty} \frac{\ln Z_N}{N} \sim \iint_{-\pi}^{\pi} \ln\left(\frac{1}{2}\left(y+\frac{1-y}{1+y}-\frac{1}{y}-\frac{1+y}{1-y}\right)-(\cos x+\cos z)\right)dxdz \tag{6}$$

where $y = e^{-k}$. Then the global analytic structure is given by the locus of points for which the argument of the log on the right hand side is zero (the circle of centre -1 and radius $\sqrt{2}$ and the circle of cnetre $+1$ and radius $\sqrt{2}$ in the complex y plane). Points corresponding to real k on these lines are the ferromagnetic and antiferromagnetic phase transition points. The zeros thus provide boundaries between the ordered and disordered phases (so that order parameters cannot be analytically continued) and may be thought of as responsible for thermodynamic critical behaviour [8]. Infact, given that the boundaries are single lines, we could have constructed this picture without solving the model. This is because the zeros of the partition function must transform into each other under duality. There is only one ferromagnetic (and one antiferromagnetic) transition and the partition function is the limit of a real polynomial. Thus the line of zeros associated with the ferromagnetic transition lies on the inversion circle of the duality transformation:

$$y \to \frac{1-y}{1+(Q-1)y} \tag{7}$$

i.e. the circle in which (6) inverts the complex plane:

$$y(\theta) = \frac{1}{1-Q}(1+\sqrt{Q}e^{i\theta}) \qquad \text{for} \quad 0 \leq \theta \leq \pi$$

The additional symmetry of the partition function under:

$$y \to \pm\frac{1}{y} \tag{8}$$

then implies the circle associated with the antiferromagnetic point.

A second (trivial) example is the $Q = 1$ case. Here there is only one configuration and all the zeros lie at $y = \infty$ ($\exp(k) = 0$). It is interesting to note that Baxter's critical points (equations (4) and (5)) for $Q = 1$ include this point and also the critical point of the percolation model (see later).

Now regarding the Ising model result as an Ice-type model with $Q = 2$ we can envisage changing Q smoothly so that the loci of zeros move towards a description of other Potts models (at least on a finite lattice). In particular we know that the ferromagnetic line intersects the real axis at the dual point and continues to be a boundary between ordered and disordered phases. This suggests that the zeros of the partition function lie on the inversion circles (for $Q = 3, 4, 5, \ldots$).

A rigorous proof of this result is not presently available. However, there is much evidence that the zeros of finite lattice partition functions accumulate on or close to the thermodynamic limit boundaries for sufficiently large lattices [7, 9, 10]. There is again no rigorous way to extrapolate from such finite lattice results. However, since we only have to recognise a simple geometrical pattern (a circle), the finite lattice result should be sufficient.

The distribution of zeros in $\exp(k)$ ($= 1/y$) for an 8×10 lattice $Q = 3$ state model are shown (for example) in figure 2. The discrepancy from a square lattice is associated with the (non trivial) construction of self-dual boundary conditions. The inversion circle for $Q = 3$ has centre $+1$ and radius $\sqrt{3}$ in the $\exp(k)$ plane and in this case the zeros associated with the ferromagnetic transition indeed lie exactly on the circle (the zeros are obtained from the exact partition function by a Newton Raphson technique iterated to within about 15 decimal places, and are well separated on this scale). We will discuss the remainder of the distribution (asociated with the antiferromagnetic transition at $\exp(k) = 0$ and its dual, $\exp(k) = -2$) later. Similar results are obtained for the other lattice sizes (presently up to 130 sites) and boundary conditions. In particular figure 3 shows the same plot for a 10×12 lattice 3-state model with periodic boundary conditions in one direction and free ($k = 0$) boundaries in the other. These boundary conditions are not self-dual, so we see that deviations due to such finite size effects are already small.

8. *Finite lattice dichromatic polynomials*

Similar results are also obtained for $Q > 3$. It is particularly instructive to consider the dichromatic polynomial (equation (2)), in which Q appears as an explicit variable, for a finite lattice (figure 1). As a polynomial in $\exp(k)$ the order is unchanged for any Q, so we may then follow the paths of the finite set of zeros. The results are algebraically complicated but there are many ways to check them (series expansions for Q,v; direct enumeration for $Q =$ integer etc.).

The dichromatic polynomial is associated with many interesting combinatorial problems. For example we observe that Baxter's equation (5) gives real solutions for negative Q. There is no equivalent Potts model and the relevant ice-type model does not have positive weights. However this mixed weight model is equivalent to a critical homogeneous model at two points. For $Q = -2$ these points are indicated – I and II – in figure 4, together with the roots of the finite lattice dichromatic polynomial. In this case (and infact for general negative Q) we see that the roots of the dichromatic polynomial seem to be accumulating on lines consistent with the critical points formally given by Baxter's critical curves (and no others).

In general the inversion circle associated with the duality transformation (7) is given by:

$$y(\theta) = \frac{1}{1-Q}(1 + |\sqrt{Q}|e^{i\theta}) \qquad \text{for} \quad 0 \leq \theta \leq \pi \tag{9}$$

with $y_{dual}(\theta) = y(\lambda - \theta)$ (where $e^{i\lambda} = Q/|Q|$). From this we find that the zeros of the $Q =$ negative polynomial do not lie on or near the inversion circle. Thus we cannot deduce the exact locus of points in this case.

For $Q > 2$ we again find evidence that the limiting distribution of zeros associated with the ferromagnetic transition lie on the inversion circle. Figure 5 shows the trajectories of zeros in $\exp(-k)$ as Q is varied from 2 to 5, and figure 6 shows $Q = 2, 3, 4, 5, 6$ for a larger lattice. Notice from equation (9) that the radius of the inversion circle decreases with Q in the $\exp(-k)$ plane, and infact the trajectories on the right hand side of figures 5 or 6 reproduce this behaviour. The finite lattice boundary conditions used in these cases are not exactly self-dual so the zeros lie close to but not on the circle (as above this could be rectified for integer Q by a more complicated set of boundary conditions). The zeros associated with the antiferromagnetic region are more sensitive to finite size effects (in particular the frustration of antiferromagnetic ground states by lattice boundaries). No geometrical shape is suggested for their distribution (and of course we have not yet identified a likely candidate on symmetry grounds).

It is interesting to note in passing that the line separation of zeros for a given Q yields an approximation to the limiting line density and hence the α critical exponent [8, 10, 11]. Our discussion of the shape of the phase boundaries has relied on the symmetries of the partition function under transformations of the global coupling constant. As such it provides no information on the line density. However by plotting the finite lattice line separation against the distance from the critical point for a sequence of different lattice sizes we find a roughly fixed power law dependence (with an overall density increase with lattice size). The result is consistent with the known α exponent for $Q = 3, 4$ [11]. However we expect severe finite size effects for $Q > 4$ since the α exponent changes discontinously in Q at $Q = 4$. This requires a "phase transition in Q" in the dichromatic polynomial, which is a thermodynamic limit effect (for finite lattices the polynomial is of finite order in Q).

For $Q = 1$ we have already remarked that the Potts model (and its transfer matrix) is trivial. However, Baxter's calculation for the critical points is sensitive to the full ice model transfer matrix (which is larger for $Q = 1$). One of the critical points obtained in this way is associated with the percolation problem critical point (equation (4)). Infact the quantity:

$$\left(\frac{\partial}{\partial Q}\ln Z(Q,k)\right)\Bigg|_{Q=1} = \frac{\sum_G C v^l}{\sum_G v^l} \tag{10}$$

which gives the average over all graphs of the number of disconnected components, is non-trivial for the finite lattice dichromatic polynomial ($p = v/(1+v)$ is the probability of a link being "occupied" in percolation terminology, see [1] and references therein). The roots of the numerator are again collected on lines which approach the real axis close to the solutions to equations (4) and (5).

4. *An interesting exercise*

The argument of the log in the Ising model partition function, equation (6), is proportional to the lowest order polynomial which has the invariance properties (up to overall factors) of the full partition function.

The fact that this is only quartic in y explains why the locus of zeros is described by the inversion circle (i.e. to satisfy the symmetries under duality and complex conjugation simultaneously). Observing that other models have this property we might conjecture that their partition functions can also be written as integrals over simple invariant polynomials. By establishing the full invariance properties of partition functions we might then construct these polynomials and hence obtain the full global analytic structure.

Such polynomials are not easy to construct because the generalised form of transformation (8) maps out of the homogeneous coupling plane. The transformation is usually thought of as a consequence of the global Z_2 symmetry of the Ising model. However it may also be obtained from the inversion properties of the transfer matrix as follows.

Baxter has shown that the eigenvalues of an appropriate Ising model transfer matrix are inverted by the transformation:

$$(b, c) \rightarrow (-c, \frac{1}{b}) \tag{11}$$

where b and c are the generalisation of $1/y$ for different horizontal and vertical bond strengths respectively (see [1] and lectures in this volume). When the Boltzmann weights are positive it is the largest eigenvalue which gives the infinite lattice partition function [12]; unfortunately the inverting trasformation breaks this condition. Repeating the trasformation to restore the original eigenvalues also restores the original weights. However, by using another symmetry of the model, that of 90 degrees rotation, between two inverting transformations, we obtain the invariance under (8). The inverting transformation may be generalised to Q-states whereupon (8) becomes:

$$(b, c) \rightarrow (\frac{1}{2-Q-b}, 2-Q-\frac{1}{c}) \tag{12}$$

This maps out of the $b = c$ plane for $Q \neq 2$. However the general (inhomogeneous) partition function must be invariant under (12) and this puts some restrictions on the form of the homogeneous lattice function. For example the lowest order invariant under all the known symmetries of the partition function is given by:

$$\left(\frac{b-t}{b-s} - t\frac{c-s}{c-t}\right)\left(\frac{c-t}{c-s} - t\frac{b-s}{b-t}\right)\Bigg|_{b=c;Q=3} \sim \left(\frac{b^2+2b}{b^2+b+1}\right)^2$$

where t and s are the fixed points of (12):

$$t, s = 1 - \frac{Q}{2} \pm \sqrt{Q(Q-4)}$$

Substituting into an expression of the form of equation (6) we obtain the following zero distribution in the complex exp (k) plane – a circle of centre $+1$ with radius $\sqrt{3}$, and a circle of centre -1 with radius 1. This should be compared with the $Q = 3$ state finite lattice result in figures 3 and 4.

Acknowledgements

I would like to thank Prof. Mario Rasetti for inviting me to Torino and for pointing out the existence of a generalised inverting transformation.

References

[1] Baxter R.J., Exactly Solved Models in Statistical Mechanics (Academic Press, London, 1982).

[2] Baxter R.J. et al, J. Phys **A8** (1976)397.

[3] Savit R., Rev. Mod. Phys. **52** (1980)453.

[4] Wu F.Y., Rev. Mod. Phys. **54** (1982)235.

[5] Temperley H.N.V. and Lieb E.H., Proc. R. Soc. London **A322** (1971)251.

[6] Baxter R.J., Proc. R. Soc. London **A383** (1982)43.

[7] Kaufman B., Phys. Rev. **76** (1949)1244.

[8] Fisher M.E., Lectures in Theoretical Physics (University of Colorado Press, Boulder, 1964)vol.7c, p.1.

[9] Abe Y. and Katsura S., Prog. Theor. Phys. **43** (1970)1402.

[10] Martin P.P., Nucl. Phys. B**225** [$FS9$] (1983)497.

[11] Martin P.P., Ice-type models, Dichromatic Polynomials ..., Queen Mary College preprint (1984)QMC-84-2 and references therein.

[12] Bellman R., Introduction to Matrix Analysis (McGraw-Hill, New York, 1960)p.278.

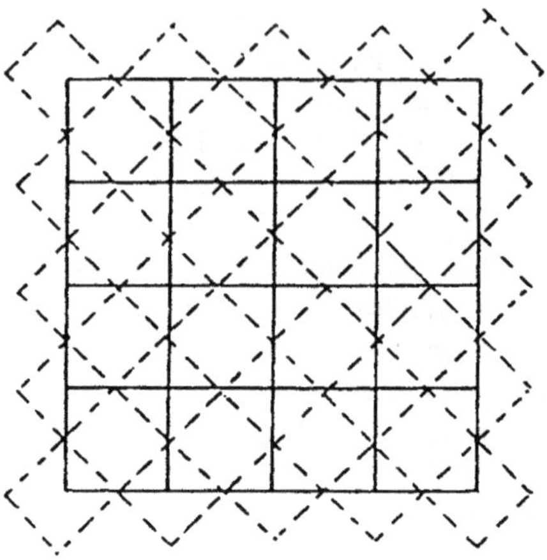

Figure 1. The medial lattice (dashed lines) associated with a 5 $\times$ 5 original lattice (solid lines).

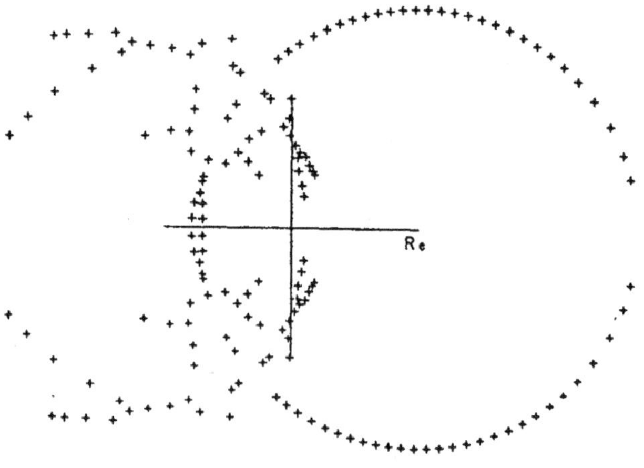

Figure 2. Zeros of the partition function plotted in the complex e^k ($=$ $1/y$) plane for the self-dual 8 $\times$ 10 lattice three state model.. The real axis (of unit length) is indicated.

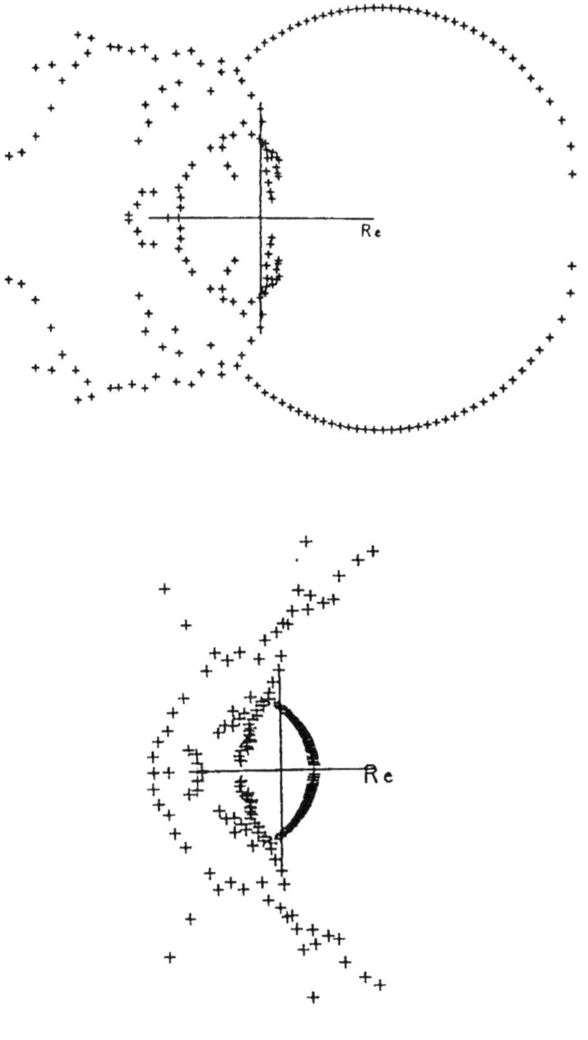

Figure 3. Zeros for the 10 $\times$ 12 lattice three state model:
a) plotted in e^k;
b) plotted in e^{-k}.

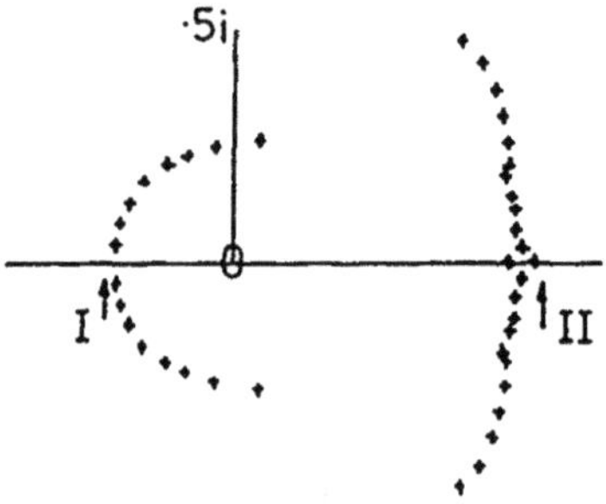

Figure 4. Zeros of the dichromatic polynomial associated with the lattice of figure 1 for $Q = -2$, plotted in y ($= e^{-k}$). Solutions to equation (5) for $Q = -2$ are also indicated (I,II).

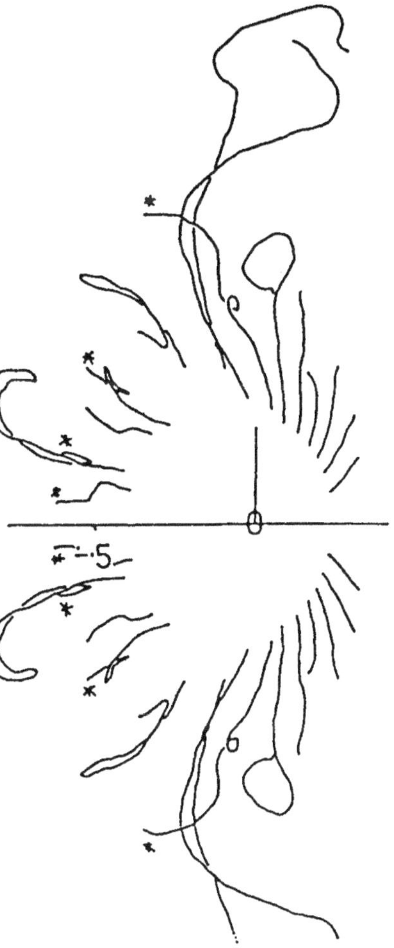

Figure 5. Trajectories of zeros of the dichromatic polynomial in e^{-k} as Q is varied from $Q = 6$ (innermost ends) to $Q = 2$ or $Q = 2.01$ (marked *, see below). The plot is obtained by considering many intermediate Q values. Since the dichromatic polynomial is of finite order in Q it is easy to see that (neglecting the vicinity of a multiple root at $Q = 2$, $e^{-k} = -1$) small changes in Q produce small changes in the positions of zeros.

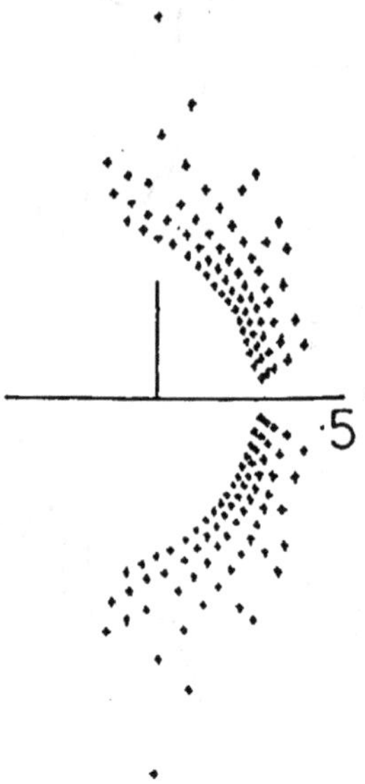

Figure 6. Zeros close to the ferromagnetic transition point for $Q = 2, 3, 4, 5, 6$ ($Q = 6$ innermost) on a 6×7 lattice, plotted in e^{-k}. For comparison with figure 2, invert in the unit circle.

Soliton equations, τ-functions and coherent states

G. M. D'ARIANO

Dipartimento di Fisica "A. Volta"
Universita' di Pavia, Italy
and

M. G. RASETTI

Dipartimento di Fisica
Politecnico di Torino, Italy

Introduction

It has been shown by M. Jimbo and T. Miwa [1] in a section of this book that the Bäcklund groups (i.e. the groups of transformations of solutions) for soliton equations are infinite dimensional Lie groups whose Lie algebras of infinitesimal generators are Kac-Moody algebras of infinite-order differential operators, called vertex operators. They have also shown that if one realizes the space of the complex polynomial algebra in terms of a Fock space of charged Fermions, writing the differential operators in terms of Fermi operators, the soliton equations become nothing but the defining differential equations of the group orbit of the highest weight vector in an infinite dimensional Fock space.

We will show that this algebraic treatment of soliton equations has a nice quantum mechanical interpretation: the solutions of the soliton equations can be viewed as quantum coherent states of an harmonic Fermi gas and the soliton dynamical evolution is thus mapped into a quantum hamiltonian evolution. The latter, which is coherence preserving, can be mapped back once more into a classical hamiltonian flow which corresponds to a succession of infinitesimal Bäcklund transformations.

Coherent states

One can define abstractly quantum coherent states associated to a Lie group $\mathcal{G}$ as follows [3, 4].

Let $\mathcal{U}$ a unitary irreducible representation of the Lie group $\mathcal{G}$ acting on the Hilbert space S of the states of the dynamical system. For every fixed "origin vector" $| \omega > \in S$ the manifold $\mathcal{M}_\omega$ of the coherent states is identified with the $\mathcal{G}$-orbit of the vector $| \omega >$ in S:

$$\mathcal{M}_\omega = U(\mathcal{G}) \mid \omega > \subset S \tag{1}$$

By definition one has:

$$\mathcal{M}_\omega \sim \mathcal{G}/K_\omega \tag{2}$$

i.e. the coherent states $| w >_\omega$ are labelled by points of the left coset space $w \in \mathcal{G}/K_\omega$, where K_ω is the stability subgroup of the vector $| \omega >$.

In general it is convenient to enlarge the unitary representation $\mathcal{U}$ to a holomorphic representation $\mathcal{T}$ of the complexified group $\mathcal{G}^c$.

If $\mathcal{M}_\omega$ is compact one has:

$$\mathcal{G}^c/K^c_\omega \sim \mathcal{G}/K_\omega \tag{3}$$

where K^c_ω is the stability subgroup of $| \omega >$ in $\mathcal{G}^c$. There follows that:

$$\mathcal{M}_\omega = \mathcal{T}(\mathcal{G}^c) \mid \omega > \subset S \tag{4}$$

Eq. (4) can also be interpreted as the definition of the coherent state manifold for the complex Lie group $\mathcal{G}^c$.

For a semisimple Lie group $\mathcal{G}^c$ the characterization of the stability subgroup is very simple if one chooses as origin vector the highest weight vector $| \lambda >$. In this case one has [5, 6]:

$$K^c_\lambda = e^{k_\lambda} \tag{5a}$$

$$k_\lambda = h \oplus g^+ \oplus o^-_\lambda \tag{5b}$$

$$o^-_\lambda = \text{span}\{x_{-\alpha} \in g_{-\alpha} \mid \alpha \in \Delta^+, (\alpha, \lambda) = 0\} \tag{5c}$$

where exp is the exponential map, h is the Cartan subalgebra of $g =$ Lie$(\mathcal{G}^c)$, $g^+ = \bigoplus_{\alpha\in\Delta^+} g_\alpha$ is the positive root-space subalgebra and o^- is the subalgebra of negative root-spaces whose roots are orthogonal to the highest weight λ (Δ^+ is the positive roots sublattice). The stability subgroup K^c_λ is thus isomorphically identified with a parabolic subgroup of $\mathcal{G}^c$ as it contains the Borel subgroup $B = \exp(h \oplus g^+)$ and coincides with the latter only when λ belongs to the interior of the dominant Weyl chamber (i.e. $(\lambda, \alpha_i) > 0 \quad \forall \alpha_i \in \Pi \equiv$ simple roots set). The above construction gives a local chart in $\mathbb{C}^{\mathcal{N}}$ for $\mathcal{M}_\lambda$:

$$| \varsigma >_\lambda = \exp\Big(\sum_{\alpha \in \Lambda_\lambda} \varsigma^\alpha x_{-\alpha} \Big) \, | \lambda > \tag{6a}$$

$$x_\alpha \in g_\alpha, \quad \Lambda_\lambda = \{ \alpha \in \Delta^+ \mid (\alpha, \lambda) > 0 \} \tag{6b}$$

$$\varsigma = \{\varsigma^\alpha\} \in \mathbb{C}^{\mathcal{N}} \qquad \mathcal{N} = |\Lambda_\lambda| \tag{6c}$$

(In the following we shall drop the index λ of the ket $| \varsigma >_\lambda$ whenever not necessary).

Thus $\mathcal{M}_\lambda$ is an almost complex manifold: indeed it is a Kaehler manifold with metric given by:

$$ds^2 = 2 \sum_{\alpha, \beta} g_{\alpha \bar{\beta}} d\varsigma^\alpha \, d\bar{\varsigma}^\beta \tag{7a}$$

$$g_{\alpha \bar{\beta}} = \frac{\partial^2 F}{\partial \varsigma^\alpha \partial \bar{\varsigma}^\beta} \tag{7b}$$

$$F(\varsigma, \bar{\varsigma}) = \ln < \varsigma \mid \varsigma > \tag{7c}$$

(The function F is positive definite as a consequence of Schwartz's inequality

$$< \varsigma \mid \varsigma > \geq |< \varsigma \mid \lambda >|^2 / < \lambda \mid \lambda > = 1).$$

The physical content of the above algebraic definition of coherent states lies mainly in their dynamical behaviour.

The quantum propagator between two (normalized) coherent states can be written as a path integral of the form [7]:

$$\begin{aligned} < \varsigma'', t'' \mid \varsigma', t' > &= < \varsigma'' \mid \exp\Big[-i\hat{H}(t'' - t')/\hbar \Big] \mid \varsigma' > = \\ &= \int D[\varsigma(t)] \exp\left\{ \frac{i}{\hbar} S \right\} \end{aligned} \tag{8}$$

where the action functional is given by:

$$S[\varsigma(t)] = \int_{t'}^{t''} L dt = \int_{t'}^{t''} < \varsigma(t) \mid i\hbar \partial_t - \hat{H} \mid \varsigma(t) > dt \tag{9}$$

with the Lagrangian:

$$L = \frac{i\hbar}{2} \sum_{\alpha\in\Lambda_\lambda} \{\dot{\varsigma}^\alpha \partial_{\varsigma^\alpha} F(\varsigma,\bar{\varsigma}) - \dot{\bar{\varsigma}}^\alpha \partial_{\bar{\varsigma}^\alpha} F(\varsigma,\bar{\varsigma})\} - H(\varsigma,\bar{\varsigma}) \qquad (10)$$

$H(\varsigma,\bar{\varsigma})$ denotes the diagonal element $< \varsigma \mid \hat{H} \mid \varsigma >$ of the system Hamiltonian $\hat{H}$. Finally the measure in (8) is:

$$\mathcal{D}[\varsigma(t)] = \prod_{t'<t<t''} d\mu(\varsigma(t)) \qquad (11)$$

$$d\mu(\varsigma) = c_\lambda \det(g_{\alpha\bar{\beta}}) \bigwedge\nolimits_{\alpha\in\Lambda_\lambda} d\varsigma^\alpha \wedge d\bar{\varsigma}^\alpha$$
$$\int_{\mathcal{M}_\lambda} d\mu(\varsigma) = 1 \qquad (12)$$

where c_λ is a normalization constant.

The stationary phase approximation ($\delta S = 0$) of eq. (8) leads to the Euler-Lagrange equations for the trajectory $\varsigma(t)$ which can be put in the Hamilton's form:

$$i\hbar \sum_{\beta\in\Lambda_\lambda} g_{\alpha\bar{\beta}} \dot{\varsigma}^\beta = \frac{\partial H}{\partial \bar{\varsigma}^\alpha}$$
$$-i\hbar \sum_{\beta\in\Lambda_\lambda} g_{\alpha\bar{\beta}} \dot{\bar{\varsigma}}^\beta = \frac{\partial H}{\partial \varsigma^\beta} \qquad (13)$$

Thus $\mathcal{M}_\lambda$ is interpreted as a curved canonical phase space for the system with metric given by (7 b). Also eq. (9) shows that the coherence preserving Schrödinger evolution of the quantum state coincides with the classical lagrangian flow on the phase space. Infact, if one computes the variation of the action one gets, after an integration by parts:

$$\delta S = \int_{t'}^{t''} \left\{ \left[\delta < \varsigma(t) \mid \right] (i\hbar\partial t - \hat{H}) \mid \varsigma(t) > + < \varsigma(t) \mid (-i\hbar\overleftarrow{\partial}_t - \hat{H}) \left[\delta \mid \varsigma(t) > \right] \right\} d \qquad (14)$$

where:

$$\delta < \varsigma(t) \mid = \sum_{k=1}^{\dim \mathcal{T}} \delta f_k(\varsigma(t)) < k \mid \qquad f_k(\varsigma) = < \varsigma \mid k > \qquad (15)$$

$\mid k >$ being a complete orthonormal set of vectors in S.

It is clear from eq. (14) that if the time dependent coherent state $|\varsigma(t)\cdot>$ satisfies the Schrödinger equation, the first variation of the action is zero. In other words, if the Schrödinger evolution, starting on the manifold $\mathcal{M}_\lambda$, remains on the manifold for every time t, the trajectory of the representative point on $\mathcal{M}_\lambda$ coincides with the classical Euler-Lagrange trajectory. Furthermore it is possible to characterize in algebraic terms the coherence preserving hamiltonian [5, 6]: it should be an element of the algebra g =Lie($\mathcal{G}$), if the latter is semisimple, whereas it can belong to a Levi extension (by a semisimple algebra) of g if this is solvable.

In summary, the given definition of coherent states permits to interpret also in the generalized case the coherent state manifold as the canonical phase space and moreover it implies that the quantum coherence-preserving time evolution coincides with the classical lagrangian one.

Summary of main results of τ-function theory

For the sake of simplicity we refer to the KP (Kadomtsev-Petviashvili) [8, 9] hierarchy, which is also the most basic.

The KP equation reads:

$$3u_{yy} - (4u_t - 6uu_x - u_{xxx})_x = 0 \tag{16}$$

The whole hierarchy involves of course an infinity of variables that we denote $z = \{z_i\}$ (with $z_1 = x$, $z_2 = y$, $z_3 = t$). Hirota's bilinearization technique [10] – upon setting $u = 2(\ln \tau)_{xx}$ – allows writing (16) in the form:

$$(D_x^4 + 3D_y^2 - 4D_xD_t)\tau \cdot \tau = 0 \tag{17}$$

where Hirota's bilinear differential operators are defined, for any polynomial P by:

$$P(D_x, D_y, D_t)f \cdot g = P(\partial_x, \partial_y, \partial_t) \cdot \left.[f(x+x', y+y', t+t')g(x-x', y-y', t-t')]\right|_{x'=y'=t'=0} \tag{18}$$

The Lie algebra of the infinitesimal Bäcklund transformation generators for eq. (7)is given by [1]:

$$\mathcal{A} = \text{span}\{Z_{ij}(z, \partial)\} \oplus \mathbb{C} \tag{19}$$

where the generating function of the differential operators $Z_{ij}(z,\partial)$ writes:

$$Z(p,q) = \frac{q}{p-q}\left\{\exp\left[\sum_{n=1}^{\infty}(p^n - q^n)z_n\right]\exp\left[-\sum_{n=1}^{\infty}\frac{1}{n}(p^{-n}-q^{-n})\frac{\partial}{\partial z_n}\right]\right\}$$
$$Z(p,q) = \sum_{i,j\in\mathbf{Z}} Z_{ij}(z,\partial)p^i q^{-j} \tag{20}$$

where $\partial = \{\partial/\partial z_n\}$. For example the transformation:

$$\tau(z) \mapsto e^{aZ(p,q)}\tau(z) \qquad\qquad a\in\mathbb{C} \tag{21}$$

is a Bäcklund transformation for the bilinear KP hierarchy which maps a solution of eq. (17) to another solution.

The Lie algebra $\mathcal{A}$ is isomorphic to $\mathcal{GL}(\infty)$, i.e. the Lie algebra of the infinite dimensional sector diagonal matrices (that is, there exists an integer N such that the matrix elements $a_{ij} = 0$ for $|i-j| > N$).

The above description of the KP τ-function can be cast into an algebraic language. Consider the "vertex operators" [11] out of which (20) is constructed:

$$X(k) = \exp\left(\sum_{n=1}^{\infty} z_n k^n\right)\exp\left(-\sum_{n=1}^{\infty}\frac{1}{n}\frac{\partial}{\partial z_n}k^{-n}\right) \tag{22}$$

and its formal adjoint:

$$\overline{X}(k) = \exp\left(-\sum_{n=1}^{\infty} z_n k^n\right)\exp\left(\sum_{n=1}^{\infty}\frac{1}{n}\frac{\partial}{\partial z_n}k^{-n}\right) \tag{23}$$

The above operators realize a correspondence between the space of the polynomial algebra $\mathbb{C}[z]$ and the Fock space $\mathcal{F}$ of charged Fermions $\{\psi_i, \overline{\psi}_i\}$, $i\in\mathbf{Z}$, by the Clifford algebra module isomorphism generated by the identification:

$$\psi_i = \hat{X}_i\,; \quad \overline{\psi}_i = \hat{\overline{X}}_i \tag{24}$$

where the $\hat{X}_i$'s and $\hat{\overline{X}}_i$'s are defined in the following way. Upon setting:

$$X(k) = \sum_{i\in\mathbf{Z}} X_i(z,\partial)k^i$$
$$\overline{X}(k) = \sum_{i\in\mathbf{Z}} \overline{X}_i(z,\partial)k^{-i} \tag{25}$$

consider copies $\{V_l\}$ of $\mathbb{C}\left[z\right]$. Then for $f_l(z) \in V_l$,

$$\begin{aligned} \hat{X}_i : V_l &\longrightarrow V_{l+1}; \qquad f_l(z) \longrightarrow X_{i-l}(z,\partial) f_l(z) \\ \hat{\overline{X}}_i : V_l &\longrightarrow V_{l-1}; \qquad f_l(z) \longrightarrow X_{i-l+1}(z,\partial) f_l(z) \end{aligned} \tag{26}$$

The Lie algebra (19) is now realized as follows:

$$\mathcal{A} = \text{span}\{:\psi_i \overline{\psi}_j:\} \oplus C \tag{27}$$

where : : denotes the usual normal ordered product defined according to Wick's theorem, the linear span is done in terms of $\mathcal{GL}(\infty)$ matrices and C is the center, spanned by the identity $\mathbf{I}$ and the non-trivial element

$$H_0 = \sum_{i \in \mathbf{Z}} :\psi_i \overline{\psi}_i: \tag{28}$$

It follows that the representation of $\mathcal{A}$ on the Fock space is reducible and the Fock space is decomposed into eigenspaces of H_0, the "charged subspaces":

$$\mathcal{F} = \oplus_n \mathcal{F}_n \tag{29}$$

Eqs. (26) define then an isomorphism between V_n and $\mathcal{F}_n$. Furthermore $\mathcal{A}$ has an Heisemberg subalgebra $\mathcal{E}$:

$$\mathcal{E} = \text{span}\{H_n, \mathbf{I}; n \in \mathbf{Z} - \{0\}\} \tag{30}$$

$$H_n = \sum_{i \in \mathbf{Z}} :\psi_i \overline{\psi}_{i+n}: \quad ; \qquad n \neq 0 \tag{31}$$

with commutation relations:

$$\left[H_n, H_m\right] = n\delta_{n,-m}\,\mathbf{I} \tag{32}$$

The existence of such a subalgebra enables to construct an explicit realization of the isomorphism between the vector spaces V_n and $\mathcal{F}_n$ and between the algebras of operators acting on them. The isomorphism is realized by the following map:

$$\begin{aligned} \mathcal{F} = \oplus_n \mathcal{F}_n &\longrightarrow V = \oplus_n V_n \\ a \mid 0 > &\longrightarrow \oplus_n < n \mid e^{H(z)} a \mid 0 > \end{aligned} \tag{33}$$

$$H(z) = \sum_{n=1}^{\infty} z_n H_n \tag{34}$$

where a is an arbitrary operator of the Clifford algebra, $| 0 >$ is the vacuum vector, $| n > \in \mathcal{F}_n$ ($| n > = \psi_{n-1} \cdots \psi_0 | 0 >, n > 0$; $| n > = \overline{\psi}_n \cdots \overline{\psi}_{-1} | 0 >, n < 0$) is the vector of charge n selected as highest weight vectors for the irreducible components of the representation of $\mathcal{A}$ on $\mathcal{F}$. We can now use the above isomorphism to write the space of τ-functions for the KP hierarchy. As the Bäcklund group is transitive on the space of solutions and $\tau = 1$ is a solution, the τ-functions manifold can be identified with the Bäcklund group orbit on the constant function $\tau = 1$. The orbit can be written in algebraic terms using the isomorphism (33) for a fixed copy V_n of the polynomial algebra as follows:

$$\tau_n(z;\gamma) = < n | e^{H(z)} \gamma | n > \qquad \gamma \in \exp(\mathcal{A}) \tag{35}$$

τ-function theory in the coherent states language

Comparing eq. (35) with the definition (4), it appears clearly that the τ-functions are nothing but coherent states representatives associated to the Lie group $\mathcal{G} = \exp(\mathcal{A})$:

$$\begin{gathered} \tau_n(z;\gamma) \equiv \Psi_n(\varsigma, z) = {}_n< \Psi | \varsigma >_n \qquad \varsigma \in \mathbb{C}^{\infty} \\ {}_n< \Psi | = < n | e^{H(z)} \\ | \varsigma >_n : \quad \text{coherent state for} \quad \mathcal{G} = \exp(\mathcal{A}) \end{gathered} \tag{36}$$

The characterization of the stability subgroup of $| n >$ is very similar to that for a semisimple Lie group. The subalgebras h, $g^{\pm}$ of $\mathcal{A} = \mathcal{GL}(\infty)$ are given, in the charged Fermion representation, by [1]:

$$\begin{aligned} g^+ &= \text{gen}\{e_i = \psi_{i-1}\overline{\psi}_i\} \\ g^- &= \text{gen}\{f_i = \psi_i\overline{\psi}_{i-1}\} \\ h &= \text{span}\{h_i = \psi_{i-1}\overline{\psi}_{i-1} - \psi_i\overline{\psi}_i\} \qquad i \in \mathbb{Z} \end{aligned} \tag{37}$$

where gen{ } denotes the algebra generated via commutations by the operators in the curly brackets. The vector $| n >$ is an highest weight vector for the irreducible component of the representation of $\mathcal{A}$ on the invariant subspace $\mathcal{F}_n$:

$$e_i | n > = 0, \qquad h_i | n > = \delta_{in} | n > \qquad \forall i \in \mathbb{Z} \tag{38}$$

The root system and the root spaces of $\mathcal{A}$, are given by [12]:

$$\Delta^+ = \{\alpha_i + \alpha_{i+1} \cdots + \alpha_j; \quad i \leq j \in \mathbf{Z}\} \qquad g_{\alpha_i} = \mathrm{span}\{e_i\} \quad (39)$$

and the stability subalgebra of $| n >$ is given by (compare (5b), (5c)):

$$k_n = h \oplus g^+ \oplus o_n^- \quad (40)$$

$$o_n = \mathrm{span}\{x_{-\alpha} \in g_{-\alpha} \mid \alpha \in \Delta^+ - \Lambda_n\} \quad (41)$$

$$\Lambda_n = \{\alpha_i + \alpha_{i+1} + \cdots + \alpha_j; \; i \leq j \in \mathbf{Z}; \; i \leq n \leq j\} \quad (42)$$

The most general coherence preserving Hamiltonian can be written as an hermitian element of $\mathcal{A}$ as follows:

$$\hat{H} = \sum_{i,j \in \mathbf{Z}} h_{ij} {:}\psi_i \bar{\psi}_j{:} \qquad h_{ij} = h_{ji}^* \in \mathbf{C} \quad (43)$$

which is the hamiltonian of an harmonic Fermi gas. The time evolution of the quantum dynamical system can be viewed as an infinite sequence of local infinitesimal contact Bäcklund transformations, in that the element $\gamma = \gamma(t) \in \mathcal{G} = \exp(\mathcal{A})$ representing the time evolution operator can be written

$$\gamma = e^{Z_1} \ldots e^{Z_k}; \quad Z_1, \ldots, Z_k \in \mathcal{A} \quad (44)$$

(The Z_i are locally nilpotent, i.e. for any $| v > \in \mathcal{F}$, one can find a sufficiently large integer M such that $Z_i^M | v >= 0$).

This procedure maps the original non-linear classical system into an equivalent quantum system.

On the other hand, under the action of the same Hamiltonian the representative point $\varsigma(t)$ of the coherent state over the phase space $\mathcal{M}_n$ evolves in time according to the lagrangian dynamics of a system of (infinitely many) canonical degrees of freedom, thus defining an hamiltonian flow – once more classical – on $\mathcal{M}_n$ itself.

In conclusion, the identification of the τ-function manifold with that of quantum coherent states, permits the mapping of the non linear soliton evolution into a quantum Schrödinger evolution and simultaneously into a classical hamiltonian flow on the coherent states manifold interpreted as a canonical phase space.

Aknowledgments

The authors aknowledge several useful discussions with M. Jimbo, T. Miwa, A.M. Perelomov and W.S. Baldoni.

References

[1] M. Jimbo and T. Miwa, in this volume (and references therein).

[2] G. D'Ariano and M. Rasetti, "Soliton equations and coherent states", Phys. Lett. A, (in press).

[3] M. Rasetti, Intl. J. Theor. Phys. 13 (1973) 425.

[4] A.M. Perelomov, Comm. Math. Phys. 26 (1972) 222.

[5] G. D'Ariano, M. Rasetti and M. Vadacchino, "Stability of Coherent States", J. Phys. A: Math. Gen. (in press).

[6] A. Cavalli, "Generalized coherent states of semisimple Lie algebras", Thesis (University of Pavia, Italy, 1984), (in italian).

[7] H. Kuratsuji and T. Suzuki, Progr. Theor. Phys. Suppl. 74 &75 (1983) 209.

[8] B.B. Kadomtsev and V.I. Petviashvili, Soviet Phys. Dokladi 15 (1970) 539

[9] V.E. Zakharov and A.B. Shabat, Functl. Anal. Appl. 8 (1974) 226

[10] R. Hirota, "Direct Method in Soliton Theory", in "Solitons", K. Bullough and P.J. Caudrey eds., Springer-Verlag, Berlin, 1980.

[11] I.B. Frenkel and V.G. Kac, Inventiones Math. 62 (1980) 23

[12] V.G. Kac, "Infinite Dimensional Lie Algebras", Birkhäuser, Boston, 1983

Commun. Math. Phys. 79, 489–505 (1981)

Communications in Mathematical Physics

Tetrahedron Equations and the Relativistic S-Matrix of Straight-Strings in $2+1$-Dimensions

A. B. Zamolodchikov

Landau Institute of Theoretical Physics, Vorobyevskoe Shosse, 2, Moscow V-334, USSR

Abstract. The quantum S-matrix theory of straight-strings (infinite one-dimensional objects like straight domain walls) in $2+1$-dimensions is considered. The S-matrix is supposed to be "purely elastic" and factorized. The tetrahedron equations (which are the factorization conditions) are investigated for the special "two-colour" model. The relativistic three-string S-matrix, which apparently satisfies this tetrahedron equation, is proposed.

1. Introduction

The progress of the last decade in studying two-dimensional exactly solvable models of quantum field theory and lattice statistical physics was motivated to some extent by using *the triangle equations.* These equations were first discovered by Yang [1]; they appeared in the problem of non-relativistic $1+1$-dimensional particles with δ-function interaction, as the self-consistency condition for Bethe's ansatz. Analogous (at least formally) relations were derived by Baxter [2], who had investigated the eight-vertex lattice model. These relations restrict the vertex weights and are of great importance for exact solvability. In particular, for the rectangular-lattice model they guarantee the commutativity of transfer-matrices with different values of the anisotropy parameter v. In the case of Baxter's general nonregular lattice $\mathscr{L}$ [3], the triangle relations for the vertex weights ensure the remarkable symmetry of the statistical system (the so-called Z-invariance): the partition function is unchanged under the deformations of the lattice, generated by the arbitrary shifts of the lattice axes. Z-invariant model on the lattice $\mathscr{L}$ is exactly solvable [3] (see also [4]).

Recently Faddeev, Sklyanin, and Takhtadjyan [5, 6] have developed a new general method of studying the exactly solvable models in $1+1$-dimensions – the quantum inverse scattering method. The triangle equations are the significant constituent of this method; they are to be satisfied by the elements of the R-matrix which determine the commutation relations between the elements of the monodromy matrix.

The triangle equations are also the central part of the theory of the relativistic purely elastic ("factorized") S-matrix in $1+1$-dimensions (for a review, see [7] and references therein). These equations (the "factorization equations") connect the elements of the two-particle S-matrix; they represent the conditions which are necessary for the factorization of the multiparticle S-matrix into two-particle ones. For the scattering theory including n different kinds of particles A_i; $i=1,2,\ldots,n$ the factorization equations have the form [7,4]

$$S^{k_1k_2}_{i_1i_2}(\theta)S^{k_3j_1}_{i_3k_1}(\theta+\theta')S^{j_2j_3}_{k_2k_3}(\theta')$$
$$=S^{k_2k_3}_{i_2i_3}(\theta')S^{k_1j_3}_{i_1k_3}(\theta+\theta')S^{j_1j_2}_{k_1k_2}(\theta), \tag{1.1}$$

where, for instance, $S^{k_1k_2}_{i_1i_2}(\theta)$ is the two-particle S-matrix, (i_1, i_2), (k_1, k_2) are the kinds of the initial (final) particles having the rapidities[1] θ_1 and θ_2, respectively; $\theta=(\theta_1-\theta_2)$. This equation has the following meaning. In the purely elastic scattering theory the three-particle S-matrix is factorized into three two-particle ones, as if the three-particle scattering were the sequence of successive pair collisions. If the rapidities $\theta_1, \theta_2, \theta_3$ of the initial particles are given, the two alternatives for the successions of these pair collisions are possible. The two different (in general) formal expressions for the three-particle S-matrix in terms of two-particle ones [the right- and left-hand sides of (1.1)] correspond to these alternatives. The conservation of the individual particle momenta requires the two "semifronts" of outgoing wave, which correspond to these two alternatives, to be coherent. The Eq. (1.1) expresses this requirement. The diagrammatic representation of the triangle Eq. (1.1) is given in Fig. 1, where the straight lines represent the "world lines" of three particles moving with the rapidities $\theta_1, \theta_2, \theta_3$. The two-particle S-matrices correspond to the intersection points of the lines; $i_a(j_a)$; $a=1,2,3$ are the kinds of the initial (final) particles; the summing over the kinds k_a of the "intermediate" particles is implied.

In [8] the version of factorized scattering theory in $2+1$-dimensions was proposed. In this theory the scattered objects are not the particles but one-dimensional formations like infinite straight-lined domain walls, which are characteristic of some models of $2+1$-dimensional field theory. We shall consider the quantum objects of this type and call them *the straight strings*. The stationary state of a moving straight string is characterized by the uniform momentum distribution along its length; its kinematics can be described completely by the direction of the string and by the transversal velocity. We assume also that the stationary states of any number of arbitrarily directed (intersecting, in general) moving straight strings are realizable[2]. The intersection points divide each string into segments, each being assumed to carry some internal quantum number i which will be called "colour". The relativistic case of the straight-string kinematics will be implied.

---

1 The rapidity of the relativistic $1+1$-dimensional particle is defined by the formulae

$$p^0_a=m\cosh\theta_a\,; \quad p^1_a=m\sinh\theta_a,$$

where p^μ_a is the two-momentum; $p^2=m^2$

2 Solutions of this type are likely in some completely integrable classical models in $2+1$-dimensions (S. Manakov, private communication)

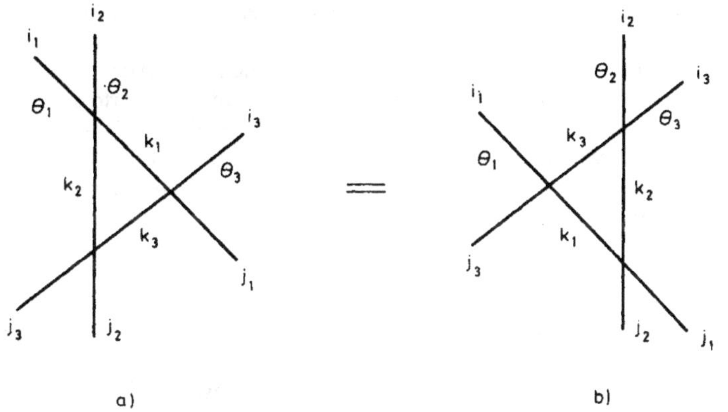

Fig. 1a and b. Diagrammatic representation of the triangle Eq. (1.1)

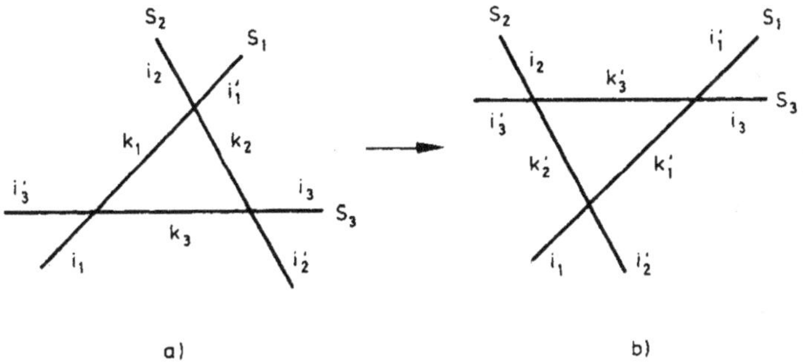

Fig. 2a and b. The initial **a** and the final **b** states of a three-string scattering

In fact, if the number L of straight-strings is less than three, the nontrivial "scattering" is impossible. The three-string scattering is "elementary". The nature of this process is illustrated in Fig. 2. The initial configuration of three-strings s_1, s_2, s_3 is shown in Fig. 2a. The indices $\{i\} = \{i_1, i_2, i_3, i'_1, i'_2, i'_3\}$ denote the colours of six "external" segments while $\{k\} = \{k_1, k_2, k_3\}$ are the colours of the "internal" ones. The motion of the strings S_a; $a = 1, 2, 3$ is such that the triangle in Fig. 2a shrinks with time. Shrinking and then "turning inside out" this triangle is the three-string scattering. After scattering, only states of the type shown in Fig. 2b appear. (This is, essentially, the meaning of the assumption of the "purely elastic" character of scattering.) The directions and velocities of the outgoing strings s_1, s_2, s_3 (Fig. 2b) coincide with those of the initial ones. The "internal" segments of strings, however, can be recoloured (in general $\{k'\} \neq \{k\}$).

In quantum theory the process shown in Fig. 2 is described by *the three-string scattering amplitude*

$$S^{i_1k_1i'_1k'_1}_{\substack{i_2k_2i'_2k'_2 \\ i_3k_3i'_3k'_3}}(\theta_1, \theta_2, \theta_3), \tag{1.2}$$

where the variables $\theta_1, \theta_2, \theta_3$ ("interfacial angles", see below) describe the scattering kinematics.

One can imagine the three-string scattering as the intersection of three planes in 2+1-dimensional space-time. These planes represent the "world sheets" swept out by the moving straight-strings. Let n_1, n_2, n_3 $[n_a^2 = (n_a^1)^2 + (n_a^2)^2 - (n_a^0)^2 = 1]$ be the normal unit vectors of the planes corresponding to the strings s_1, s_2, s_3, respectively. The mutual orientation of three planes, and, hence, the kinematics of the three-string scattering, is described completely by three invariants

$$n_1 n_2 = -\cos\theta_3 ; \quad n_1 n_3 = -\cos\theta_2 ; \quad n_2 n_3 = -\cos\theta_1 . \tag{1.3}$$

The two-plane intersection lines divide every plane s_a $(a = 1, 2, 3)$ into four parts which will be called plaquettes. The colours of string segments, denoted by the indices i_a, k_a, i'_a, k'_a in (1.2) can be obviously attached to twelve plaquettes joined to the three-plane intersection point. In what follows this point will be called the vertex while the angles $\theta_1, \theta_2, \theta_3$, defined by (1.3) – the vertex variables.

The L-string scattering for $L > 3$ has similar properties: the directions and velocities of all the strings s_a; $a = 1, 2, \ldots, L$ remain unchanged after the scattering, the "internal" segments being, in general, recoloured. We assume the factorization of the multistring S-matrix: the L-string S-matrix is the product of $L(L-1)(L-2)/6$ three-string ones (1, 2), according to the idea that the L-string scattering can be thought of as the sequence of three-string collisions. The succession of this three-string collision is not determined uniquely by the directions and velocities of all the strings s_a but depends also on their "initial positions". Like the 1+1-dimensional case, the self-consistency of the factorization condition for the straight-string S-matrix requires the equality of different formal expressions for the L-string S-matrix in terms of three-string amplitudes, corresponding to the different successions of three-string collisions. It is easy to note that this requirement is equivalent to *the tetrahedron equation* shown in Fig. 3. In this figure the "world planes" of four strings s_a, $a = 1, 2, 3, 4$ (undergoing the four-string scattering) are shown. These planes form the tetrahedron in 2+1 space-time. The vertices of the tetrahedron represent the "elementary" three-string collisions; the corresponding S-matrices (1.2) are the multipliers in the expression for the 4-string S-matrix. The tetrahedra shown in Fig. 3a and 3b (which differ from each other by some parallel shift of the planes s_a) represent two possible successions of three-string collisions constituting the same four-string scattering process. The colours of the "external" plaquettes are fixed and respectively equal in the right- and the left-hand sides of the equality in Fig. 3; the summing over all possible colourings of the "internal" plaquettes (which are the faces of the tetrahedra) is implied. This tetrahedron equation should be satisfied at any mutual orientations of the planes s_1, s_2, s_3, s_4.

The 1+1-dimensional factorized S-matrix can be interpreted, after euclidean continuation, as the Z-invariant statistical model on the planar Baxter's lattice $\mathscr{L}$ (see [4]). The 2+1-dimensional factorized S-matrix of straight-strings admits similar interpretation [8]. The natural three-dimensional analog of Baxter's lattice $\mathscr{L}$ is the lattice formed by a large number L of arbitrarily directed intersecting planes in three-dimensional euclidean space. The fluctuating variables ("colours")

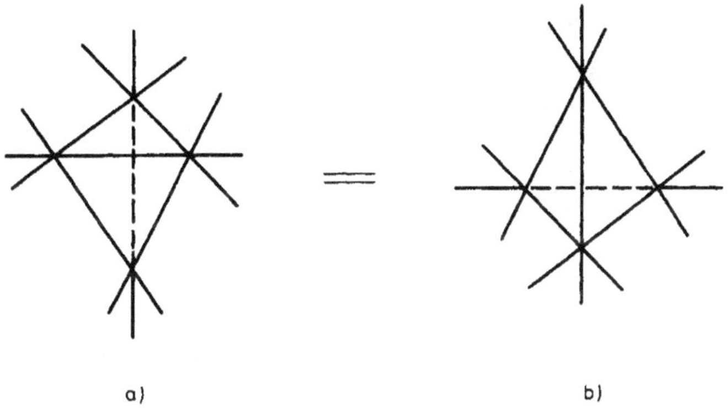

Fig. 3a and b. Diagrammatic representation of the tetrahedron equations

are attached to the lattice plaquettes. The partition function is defined as the sum over all possible colourings of all the plaquettes, each colour configuration being taken with the weight equal to the product of the vertex weights over all the vertices (the vertices are the points of triple intersections of the planes). The vertex weights are assumed to be the functions (common for all the vertices) of the mutual orientation of three planes intersecting in a given vertex. Identifying the vertex weights with the elements of the three-string S-matrix (1.2) (continued to the euclidean domain), one can note that, due to the tetrahedron equation (Fig. 3), the statistical system thus defined possesses Z-invariance.

The tetrahedron equation (Fig. 3) turns out to be highly overdefined system of functional equations; even in the simplest models the independent equations outnumber (by several hundredfold) the independent elements of the three-string S-matrix (1.2). Therefore, the compatibility of these equations is extremely crucial for the scattering theory of straight-strings. In [8] the *two-colour model* of straight-string scattering theory was proposed, and the explicit solution of the corresponding tetrahedron equations was found in the special "static limit" which corresponds to the case $v_a \to 0$, where v_a are the velocities of all the strings. In this paper we construct the relativistic three-string S-matrix for the two-colour model, which is apparently the solution of the "complete" tetrahedron equations. Although the complete evidence of the last statement is unknown we present some nontrivial checks.

The qualitative aspects of the factorized straight-string scattering theory have been described briefly in this Introduction; more detailed discussion can be found in [8]. In Sect. 2 the formulation of the two-colour model is given for the relativistic case. The corresponding tetrahedron equations are discussed in Sect. 3. In Sect. 4 the explicit formulae for the elements of the three-string S-matrix are proposed and the arguments that this S-matrix satisfies the tetrahedron equations are presented. In Sect. 5 it is shown that the obtained S-matrix is in agreement with the unitarity condition for the straight-string S-matrix.

2. Two-colour Model of Straight-strings Scattering Theory

Consider the relativistic scattering theory of straight-strings (see the Introduction) in which the strings' segments can carry only two colours – "white" or "black". Further, let us allow only the states satisfying the following requirement: the even number (i.e., 0, 2 or 4) of black segments can join in each point of two-string intersection. In other words, in any allowed state the black segments form continuous polygonal lines (which may intersect) without ends. Certainly, all the elements of the three-string S-matrix converting the allowed states into unallowed ones (and vice versa) are implied to be zero.

As it is explained in the Introduction, the three-string scattering kinematics can be represented by means of three intersecting "world planes" s_1, s_2, s_3 in $2+1$ space time, the vertex being the "place of collision". In the two-colour model each of the twelve plaquettes joining the vertex can be coloured into black or white so that the black plaquettes form the continuous broken surfaces without boundaries.

Each allowed coluring of these twelve plaquettes corresponds to some nonvanishing element of the three-string S-matrix.

It is convenient to perform the considerations in terms of the euclidean space-time: the "world planes" s_a can be treated as imbedded in the 3-dimensional euclidean space; each of the variables θ, defined by (1.3) being some interfacial angle. The "physical" amplitudes of scattering in the Minkowski space-time can be obtained from the euclidean formulae by means of analytical continuation.

Let us picture the "colour configuration" of the twelve plaquettes joining the vertex as follows. Consider the sphere with the vertex as its centre. The planes s_1, s_2, s_3 draw three great circles on this sphere; the variables $\theta_1, \theta_2, \theta_3$ [see (1.3)] are exactly the intersection angles of these circles. The intersection points divide each of the circles into four segments; the colours of the plaquettes can be obviously attached to these segments. Performing the stereographic projection one can map these three circles on the plane as shown in Fig. 4. This picture can be interpreted as follows. The spherical triangle I_1 in Fig. 4 corresponds to the triangle in Fig. 2a and represents the initial state of some three-string scattering process. The final state of this process (shown in Fig. 2b) is represented by the spherical triangle F_1. The variables $\theta_1, \theta_2, \theta_3$ are the interior angles of the triangles I_1 and F_1 (obviously, these triangles are equal on the sphere). Alternatively, one could consider, for instance, the triangle I_2 as representing the initial (and F_2 as the final) state of some other three-string process. This is just the cross-channel. Evidently, the transfer to this cross-channel is associated with the variable transformation

$$\theta_1 \to \pi - \theta_1; \quad \theta_2 \to \pi - \theta_2; \quad \theta_3 \to \theta_3. \tag{2.1}$$

As it is clear from Fig. 4, each three-string scattering process has four cross-channels $I_1 \to F_1$, $I_2 \to F_2$, $I_3 \to F_3$, $I_4 \to F_4$.

We shall assume the P and T invariances of the straight-string scattering theory [8], and also its symmetry under the simultaneous recolouring of all black segments into white and vice versa ("colour symmetry"). Then the three-string S-matrix contains 8 independent amplitudes which are shown (together with the adopted notations) in Fig. 5.

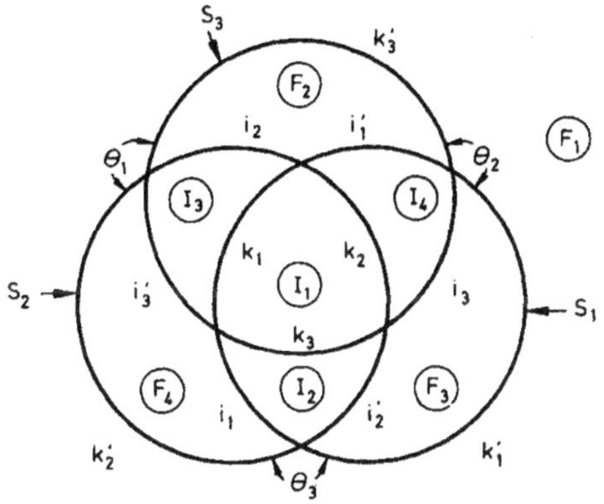

Fig. 4. Stereographic projection of the sphere surrounding the vertex

a) $S(\theta_1,\theta_2,\theta_3)$ b) $a(\theta_1,\theta_2,\theta_3)$ c) $V(\theta_1,\theta_2,\theta_3)$

d) $R(\theta_1,\theta_2,\theta_3)$ e) $K(\theta_1,\theta_2,\theta_3)$ f) $w(\theta_1,\theta_2,\theta_3)$

g) $\sigma(\theta_1,\theta_2,\theta_3)$ h) $T(\theta_1,\theta_2,\theta_3)$

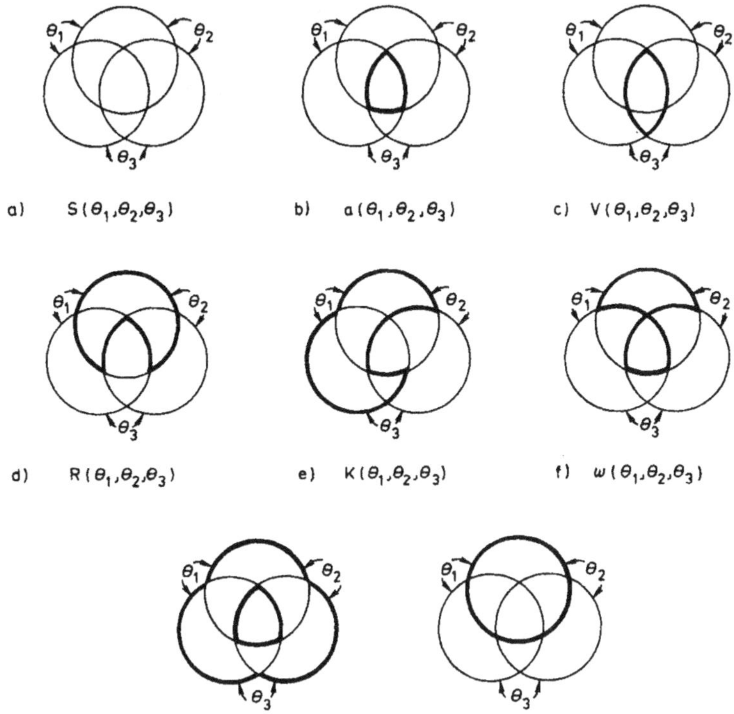

Fig. 5a–h. Eight "colour configurations" of twelve plaquettes joining the vertex, and the notations for corresponding three-string scattering amplitudes. The white (black) circular segments are represented by the ordinary (solid) lines

It is convenient to introduce, apart from the amplitudes defined in Fig. 5 the following 5 functions

$$U(\theta_1,\theta_2,\theta_3)=\mathcal{a}(\pi-\theta_1,\pi-\theta_2,\theta_3), \tag{2.2a}$$

$$L(\theta_1,\theta_2,\theta_3)=V(\pi-\theta_1,\theta_2,\pi-\theta_3), \tag{2.2b}$$

$$\Omega(\theta_1,\theta_2,\theta_3)=R(\pi-\theta_1,\pi-\theta_2,\theta_3), \tag{2.2c}$$

$$H(\theta_1,\theta_2,\theta_3)=R(\pi-\theta_1,\theta_3,\pi-\theta_2), \tag{2.2d}$$

$$W(\theta_1,\theta_2,\theta_3)=\sigma(\pi-\theta_1,\pi-\theta_2,\theta_3), \tag{2.2e}$$

which describe the cross-channels of the processes shown in Fig. 5. The three-string amplitudes should possess the following symmetries, which are the consequences of P, T, "colour" and crossing symmetries

$$S(\theta_1,\theta_2,\theta_3)=S(\theta_2,\theta_1,\theta_3)=S(\theta_1,\theta_3,\theta_2)$$
$$=S(\pi-\theta_1,\pi-\theta_2,\theta_3). \tag{2.3a}$$

$$\mathcal{a}(\theta_1,\theta_2,\theta_3)=\mathcal{a}(\theta_2,\theta_1,\theta_3)=\mathcal{a}(\theta_1,\theta_3,\theta_2),$$
$$U(\theta_1,\theta_2,\theta_3)=U(\theta_2,\theta_1,\theta_3). \tag{2.3b}$$

$$V(\theta_1,\theta_2,\theta_3)=V(\theta_2,\theta_1,\theta_3)=V(\pi-\theta_1,\pi-\theta_2,\theta_3),$$
$$L(\theta_1,\theta_2,\theta_3)=L(\theta_2,\theta_1,\theta_3). \tag{2.3c}$$

$$R(\theta_1,\theta_2,\theta_3)=R(\theta_2,\theta_1,\theta_3), \quad H(\theta_1,\theta_2,\theta_3)=H(\theta_2,\theta_1,\theta_3),$$
$$\Omega(\theta_1,\theta_2,\theta_3)=\Omega(\theta_2,\theta_1,\theta_3). \tag{2.3d}$$

$$\omega(\theta_1,\theta_2,\theta_3)=\omega(\theta_2,\theta_1,\theta_3)=\omega(\pi-\theta_1,\pi-\theta_2,\theta_3)$$
$$=\omega(\pi-\theta_1,\theta_2,\pi-\theta_3). \tag{2.3e}$$

$$K(\theta_1,\theta_2,\theta_3)=K(\theta_2,\theta_1,\theta_3)=K(\pi-\theta_2,\pi-\theta_1,\theta_3)$$
$$=K(\pi-\theta_1,\theta_2,\pi-\theta_3). \tag{2.3f}$$

$$\sigma(\theta_1,\theta_2,\theta_3)=\sigma(\theta_2,\theta_1,\theta_3)=\sigma(\theta_1,\theta_3,\theta_2),$$
$$W(\theta_1,\theta_2,\theta_3)=W(\theta_2,\theta_1,\theta_3). \tag{2.3g}$$

$$T(\theta_1,\theta_2,\theta_3)=T(\theta_2,\theta_1,\theta_3)=T(\pi-\theta_1,\pi-\theta_2,\theta_3)$$
$$=T(\pi-\theta_1,\theta_2,\pi-\theta_3). \tag{2.3h}$$

The analytic properties of the three-string amplitudes will be considered in Sects. 3 and 4.

3. The Tetrahedron Equations

The hardest restrictions for the three-string S-matrix come from the tetrahedron equations, which are shown schematically in Fig. 3. Here we shall choose the four "world planes" s_1, s_2, s_3, s_4 shown in this figure to be placed into the euclidean space (see Sect. 2). The three-string S-matrices associated with the vertices of the tetrahedra in Fig. 3 are the functions of corresponding vertex variables. In the two-

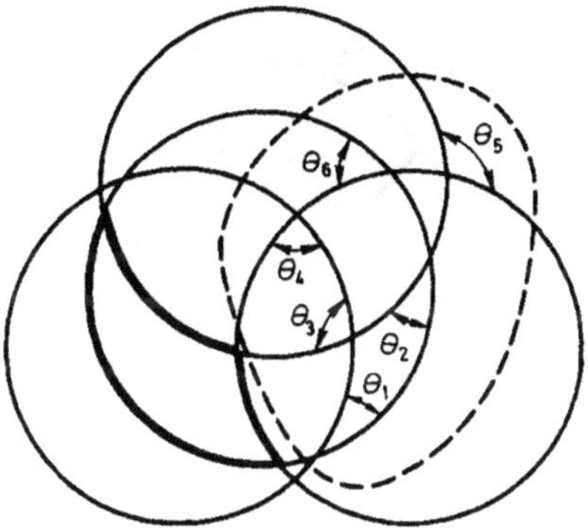

Fig. 6. Stereographic projection of large sphere surrounding the tetrahedron

colour model each plaquette in Fig. 3 can be black or white. Recall that the colours of 24 "external" plaquettes are fixed and equal in the right- and left-hand sides of the tetrahedron equation while the independent summing is performed over all colourings of "internal" plaquettes. Obviously, each allowed colouring of the "external" plaquettes gives rise to some functional equation connecting the three-string amplitudes.

To describe the colourings of the "external" plaquettes it is convenient to introduce again the large sphere (its radius is much larger than the size of the tetrahedra), taking some point near the vertices as the centre, and consider 4 great circles on the sphere corresponding to the planes s_1, s_2, s_3, s_4 (certainly, the tetrahedra in Figs. 3a and 3b are indistinguishable from this point of view). Stereographic projection of this sphere is shown in Fig. 6. The angles θ_1, θ_2, θ_3, θ_4, θ_5, θ_6, shown in this figure are just the interior interfacial angles (i.e., the angles between the planes s_a) of the tetrahedra.

Any allowed colouring of the "external" plaquettes in Fig. 3 corresponds in obvious manner to some allowed colouring of the 24 circular segments in Fig. 6 into black and white. In Fig. 6 some colouring of this type is shown as an example. This colouring gives rise, as it is evident from simple consideration, to the following functional equation

$$\begin{aligned}
&S(\theta_1,\theta_2,\theta_3)S(\theta_1,\theta_4,\theta_6)S(\theta_5,\theta_4,\theta_3)a(\theta_2,\theta_5,\theta_6)\\
&+a(\theta_1,\theta_2,\theta_3)a(\theta_1,\theta_4,\theta_6)a(\theta_4,\theta_3,\theta_5)\sigma(\theta_2,\theta_5,\theta_6)\\
&=U(\theta_1,\theta_3,\theta_2)U(\theta_1,\theta_4,\theta_6)U(\theta_4,\theta_3,\theta_5)S(\theta_2,\theta_5,\theta_6)\\
&+V(\theta_1,\theta_3,\theta_2)V(\theta_1,\theta_4,\theta_6)V(\theta_3,\theta_4,\theta_5)a(\theta_2,\theta_5,\theta_6)\,. \qquad (3.1)
\end{aligned}$$

The equation (3.1) is only one representative of the system of functional tetrahedron equations which arises if one considers all possible allowed colourings of the circular segments in Fig. 6. This system includes hundreds of independent equations and we are not able to present it here; the equation (3.1) is written down mainly for illustration.

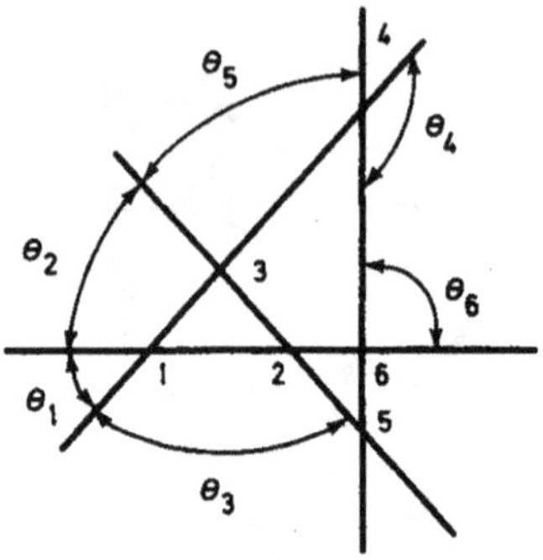

Fig. 7. Fragment of the diagram in Fig. 6 which is enclosed with the dotted curve

It is essential that the variables $\theta_1, \theta_2, \theta_3, \theta_4, \theta_5, \theta_6$ in the tetrahedron equations are not completely independent. Since the mutual orientation of four planes in three-dimensional space is determined completely by only five parameters, there is one relation between these six angles[3]. This relation can be derived, for instance, from the spherical trigonometry. To do so, concentrate on the fragment of Fig. 6, surrounded by dotted curve. This fragment is shown separately in Fig. 7, where the circular arcs are drawn schematically as the straight lines. Using the formulae of spherical trigonometry one can express the lengths of segments ℓ_{32} and ℓ_{25} in terms of the interior angles of the spherical triangles (123) and (256), respectively. On the other hand, the length of the segment $\ell_{35}=\ell_{32}+\ell_{25}$ may be expressed independently in terms of the interior angles of the triangle (354). This allows one to write

$$\left[\cos\frac{\theta_1+\theta_3-\theta_2}{2}\cos\frac{\theta_1+\theta_2-\theta_3}{2}\cos\frac{\theta_6+\theta_5-\theta_2}{2}\cos\frac{\theta_6+\theta_2-\theta_5}{2}\right]^{1/2}$$
$$-\left[\cos\frac{\theta_1+\theta_2+\theta_3}{2}\cos\frac{\theta_2+\theta_3-\theta_1}{2}\cos\frac{\theta_2+\theta_5+\theta_6}{2}\cos\frac{\theta_2+\theta_5-\theta_6}{2}\right]^{1/2}$$
$$=\sin\theta_2\left[\cos\frac{\theta_3+\theta_5-\theta_4}{2}\cos\frac{2\pi-\theta_3-\theta_4-\theta_5}{2}\right]^{1/2}. \tag{3.2}$$

Equation (3.2) is a variant of the desired relation.

4. The Solution of the Tetrahedron Equations

The relation (3.2) connecting the interior angles of the tetrahedron essentially complicates the direct investigation of the tetrahedron equations. However, one can concentrate at first on the special limiting case. Namely, consider the variables

3 Certainly, this relation is the imbedding condition of four vectors n_1, n_2, n_3, n_4 into the three-dimensional space. Its general form is $\det|n_a^\mu|=0$, where four vectors n_a^μ are treated formally as four-dimensional; $a, \mu=1,2,3,4$

$\theta_1, \theta_2, \theta_3, \theta_4, \theta_5, \theta_6$ satisfying the relation

$$\theta_1+\theta_2+\theta_3=\theta_2+\theta_5+\theta_6=\theta_4+\theta_3-\theta_5=\pi, \tag{4.1}$$

which corresponds to the limit of coplanar vectors n_1, n_2, n_3, n_4. In this case all the spherical triangles in Fig. 7 can be treated as planar ones and the relation (3.2) is certainly satisfied. From the viewpoint of straight-strings kinematics, the relation (4.1) corresponds to the limit of "infinitely slow" strings; therefore we call this case the "static limit". In the static limit the variables θ_1, θ_2, θ_3 (which are the arguments of the three-string amplitudes) are just the planar angles between the directions of three strings s_1, s_2, s_3. They satisfy the relation $\theta_1+\theta_2+\theta_3=\pi$. Hence the "static" three-string amplitudes are the functions, not of three, but of two variables, θ_1, θ_2.

Most of the tetrahedron equations do not become identities in the static limit [as it happens for the Eq. (3.1)]. Actually, considering the static limit, the number of the independent tetrahedron equations even increase, since the different cross-channels of the same "complete" equations give rise to the different "static" tetrahedron equations.

In [8] the solution of the static-limit tetrahedron equation was constructed; it has the form

$$\begin{aligned}
&S^{st}(\theta_1,\theta_2)=\sigma^{st}(\theta_1,\theta_2)=T^{st}(\theta_1,\theta_2)=W^{st}(\theta_1,\theta_2)=1;\\
&\mathcal{A}^{st}(\theta_1,\theta_2)=R^{st}(\theta_1,\theta_2)=0;\\
&L^{st}(\theta_1,\theta_2)=\omega^{st}(\theta_1,\theta_2)=-K^{st}(\theta_1,\theta_2)=\varepsilon_1 V^{st}(\theta_1,\theta_2)=\left[\operatorname{tg}\frac{\theta_1}{2}\operatorname{tg}\frac{\theta_2}{2}\right]^{1/2};\\
&H^{st}(\theta_1,\theta_2)=U^{st}(\theta_1,\theta_2)=-\varepsilon_1\Omega^{st}(\theta_1,\theta_2)=\varepsilon_2\left[\frac{\cos\left(\frac{\theta_1}{2}+\frac{\theta_2}{2}\right)}{\cos\frac{\theta_1}{2}\cos\frac{\theta_2}{2}}\right]^{1/2},
\end{aligned} \tag{4.2}$$

where the notations for the three-string amplitudes are the same as in Fig. 5 and in (2.2), the θ_3 being set equal to $\pi-\theta_1-\theta_2$; for instance,

$$U^{st}(\theta_1,\theta_2)\equiv U(\theta_1,\theta_2,\pi-\theta_1-\theta_2).$$

In (4.2) ε_1 and ε_2 are arbitrary signs; $\varepsilon_1^2=\varepsilon_2^2=1$. Expressions (4.2) satisfy all the "static" tetrahedron equations. We do not insist that (4.2) is the general solution; rather we think that it is not so.

Let us search for the solution of the "complete" tetrahedron equations which corresponds to the static limit (4.2). First consider the power expansion around the static limit. Namely, let the velocities of the scattered strings be not exactly zero but small. In this case the three-string scattering amplitudes can be conveniently considered as the functions of two angles θ_1, θ_2 (which determine the space directions of the strings s_1, s_2, s_3, see Fig. 2) and "symmetrical velocity" $w=\frac{1}{2}dr/dt$ where r is the radius of the circle inscribed in the triangle in Fig. 2. At small velocities of the strings s_1, s_2, s_3, s_4 the nonrelativistic kinematics is valid, and the "velocities" w, corresponding to four triangles (123), (256), (146), (453) in Fig. 7, are connected as follows:

$$w_{146}\frac{\sin\frac{\theta_{125}}{2}\sin\frac{\theta_2}{2}}{\sin\frac{\theta_1}{2}\sin\frac{\theta_{25}}{2}}=w_{123}\frac{\theta_{12}}{\sin\frac{\theta_1}{2}}+w_{256}\frac{\cos\frac{\theta_5}{2}}{\cos\frac{\theta_{25}}{2}};$$

$$w_{453}\frac{\sin\frac{\theta_{125}}{2}\sin\frac{\theta_2}{2}}{\sin\frac{\theta_5}{2}\sin\frac{\theta_{12}}{2}}=w_{123}\frac{\cos\frac{\theta_1}{2}}{\cos\frac{\theta_{12}}{2}}+w_{256}\frac{\sin\frac{\theta_{25}}{2}}{\sin\frac{\theta_5}{2}}, \tag{4.3}$$

where the notations $\theta_{12}=\theta_1+\theta_2$; $\theta_{25}=\theta_2+\theta_5$; $\theta_{125}=\theta_1+\theta_2+\theta_5$ are used. The investigation of the tetrahedron equations in the linear approximation (in w) leads to the result

$$\mathcal{a}(\theta_1,\theta_2,w)=-\varepsilon_1 R(\theta_1,\theta_2,w)=\varepsilon_2\lambda w\left[\sin\frac{\theta_1}{2}\sin\frac{\theta_2}{2}\sin\frac{\theta_3}{2}\right]^{-1/2}+O(w^2), \tag{4.4a}$$

$$S(\theta_1,\theta_2,w)=T(\theta_1,\theta_2,w)=1-\lambda w+O(w^2), \tag{4.4b}$$

$$\sigma(\theta_1,\theta_2,w)=W(\theta_1,\theta_2,w)=1+\lambda w+O(w^2), \tag{4.4c}$$

$$\begin{aligned}U(\theta_1,\theta_2,w)&=H(\theta_1,\theta_2,w)=-\varepsilon_1\Omega(\theta_1,\theta_2,w)\\&=U^{st}(\theta_1,\theta_2)(1+O(w^2)),\end{aligned} \tag{4.4d}$$

$$\begin{aligned}\omega(\theta_1,\theta_2,w)&=-K(\theta_1,\theta_2,w)\\&=L^{st}(\theta_1,\theta_2)(1+\lambda w\operatorname{ctg}\frac{\theta_1}{2}\operatorname{ctg}\frac{\theta_2}{2}+O(w^2))\end{aligned} \tag{4.4e}$$

$$\begin{aligned}L(\theta_1,\theta_2,w)&=\varepsilon_1 V(\theta_1,\theta_2,w)\\&=L^{st}(\theta_1,\theta_2)(1-\lambda w\operatorname{ctg}\frac{\theta_1}{2}\operatorname{ctg}\frac{\theta_2}{2}+O(w^2)),\end{aligned} \tag{4.4f}$$

where $\theta_3=\pi-\theta_1-\theta_2$, L^{st} and U^{st} are given by Eqs. (4.2), and λ is an arbitrary constant.

In studying the complete relativistic tetrahedron equations it is convenient to introduce the variables (spherical excesses)

$$\begin{aligned}2\alpha&=\theta_1+\theta_2+\theta_3-\pi,\\2\beta&=\pi+\theta_3-\theta_1-\theta_2,\\2\gamma&=\pi+\theta_1-\theta_2-\theta_3,\\2\delta&=\pi+\theta_2-\theta_1-\theta_3,\end{aligned} \tag{4.5}$$

obeying the relation

$$\alpha+\beta+\gamma+\delta=\pi. \tag{4.6}$$

Any transmutations of θ_1, θ_2, θ_3, and also any crossing transformations of the type of (2.1) lead, as one can easily verify, to some transmutations among the variables α, β, γ, δ. In fact, the quantity 2α is the area of the spherical triangle I_1 (and F_1) in Fig. 4, while 2β, 2γ, 2δ are the areas of the triangles I_2, I_3, I_4, respectively. Therefore the four cross-channels $I_1 \to F_1$, $I_2 \to F_2$, $I_3 \to F_3$, $I_4 \to F_4$ of the three-string scattering will be called α, β, γ, δ-channels, respectively.

In the static limit $\theta_1 + \theta_2 + \theta_3 \to \pi$, and we have

$$\alpha \to 0; \quad \beta \to \theta_3; \quad \gamma \to \theta_1; \quad \delta \to \theta_2. \tag{4.7}$$

The following relation is valid up to the main order in w

$$w = \sqrt{\frac{\alpha}{2}} \left[\operatorname{tg}\frac{\theta_1}{2} \operatorname{tg}\frac{\theta_2}{2} \operatorname{tg}\frac{\theta_3}{2} \right]^{1/2}. \tag{4.8}$$

Therefore, as it is seen from (4.4), the three-string amplitudes have the square-root branching plane $\alpha = 0$, which will be called the α-channel threshold. The crossing symmetry requires the amplitudes to possess the branching planes (also square-root) $\beta = 0$; $\gamma = 0$; $\delta = 0$, which are the thresholds of the β, γ, δ-channels.

These reasons allow one to write down the following formulae

$$\mathcal{U}(\theta_1, \theta_2, \theta_3) = R(\theta_1, \theta_2, \theta_3) = \varepsilon_2 \left[\frac{\sin\frac{\alpha}{2}}{\cos\frac{\beta}{2} \cos\frac{\gamma}{2} \cos\frac{\delta}{2}} \right]^{1/2}; \tag{4.9a}$$

$$-V(\theta_1, \theta_2, \theta_3) = \left[\operatorname{tg}\frac{\gamma}{2} \operatorname{tg}\frac{\delta}{2} \right]^{1/2} - \left[\operatorname{tg}\frac{\alpha}{2} \operatorname{tg}\frac{\beta}{2} \right]^{1/2}; \tag{4.9b}$$

$$\mathcal{W}(\theta_1, \theta_2, \theta_3) = -K(\theta_1, \theta_2, \theta_3) = \left[\operatorname{tg}\frac{\gamma}{2} \operatorname{tg}\frac{\delta}{2} \right]^{1/2} + \left[\operatorname{tg}\frac{\alpha}{2} \operatorname{tg}\frac{\beta}{2} \right]^{1/2}; \tag{4.9c}$$

$$S(\theta_1, \theta_2, \theta_3) = T(\theta_1, \theta_2, \theta_3) = 1 - \left[\operatorname{tg}\frac{\alpha}{2} \operatorname{tg}\frac{\beta}{2} \operatorname{tg}\frac{\gamma}{2} \operatorname{tg}\frac{\delta}{2} \right]^{1/2}; \tag{4.9d}$$

$$\sigma(\theta_1, \theta_2, \theta_3) = 1 + \left[\operatorname{tg}\frac{\alpha}{2} \operatorname{tg}\frac{\beta}{2} \operatorname{tg}\frac{\gamma}{2} \operatorname{tg}\frac{\delta}{2} \right]^{1/2}, \tag{4.9e}$$

which are in accordance with the expansion (4.4) provided $\varepsilon_1 = -1$ and $\lambda = 1$, and entirely satisfy the crossing relations (2.3). Therefore we suppose that the expressions (4.9) give the exact solution of the "complete" tetrahedron equations for the two-colour model.

Unfortunately, rigorous verification of this supposition is rather difficult. Direct substitution of (4.9) into the tetrahedron equation is complicated because of the relation (3.2), not to speak of the large number of equations to be verified. However, we have performed some simplified verifications; an example is given in the Appendix. Moreover, our supposition has been confirmed in various numerical checks.

Note that, like the triangle equations (1.1), the tetrahedron equations are homogeneous; the three-string S-matrix is determined by the equations only up to the overall factor which can be a function of the variables θ. The formulae (4.9) should be considered as expressions giving the ratios of different elements of the three-string S-matrix; the right-hand sides of all the equalities (4.9) are implied to be multiplied by some function

$$[Z(\alpha, \beta, \gamma, \delta)]^{-1}, \tag{4.10}$$

which is symmetric under arbitrary transmutations of the variables $\alpha, \beta, \gamma, \delta$ [this is forced by the crossing symmetry requirements (2.3)]. This function will be determined by the unitarity condition for the straight-strings S-matrix, studied in the next section.

5. Unitarity Condition

In the euclidean domain the variables $\alpha, \beta, \gamma, \delta$ [connected by (4.6)] are real and non-negative, and all the amplitudes (4.9) are real. The "physical" scattering of the strings s_a in Minkowski space-time corresponds to real negative values of α (provided the velocities of the strings s_a are not too large[4]. Here the amplitudes acquire the imaginary parts. Let us introduce the cutting hyperplane $\mathrm{Im}\,\alpha=\mathrm{Im}\,\beta=\mathrm{Im}\,\gamma=\mathrm{Im}\,\delta=0$; $\mathrm{Re}\,\alpha<0$ (corresponding to the branching plane $\alpha=0$) in the three-dimensional complex space of the variables $\alpha, \beta, \gamma, \delta$. Then the "upper" edge ($\mathrm{Im}\,\alpha=+0$; $\mathrm{Re}\,\beta>0$; $\mathrm{Re}\,\gamma>0$; $\mathrm{Re}\,\delta>0$) of this hyperplane represents the "physical" domain of α-channel. Continuing some amplitude to the "lower" edge $\mathrm{Im}\,\alpha=-0$, one obtains the complex-conjugated amplitude of reversed process (here we imply the T-invariance so that the amplitudes of direct and reversed processes are equal).

In the physical domain of α-channel the three-string unitarity condition should be satisfied, i.e.,

$$\sum_{k_1'k_2'k_3'} S^{i_1k_1i_1'k_1'}_{i_2k_2i_2'k_2'\,i_3k_3i_3'k_3'}(\theta_1,\theta_2,\theta_3)\,S^{i_1\ell_1i_1'k_1'}_{i_2\ell_2i_2'k_2'\,i_3\ell_3i_3'k_3'}(\theta_1,\theta_2,\theta_3)^*=\delta^{\ell_1}_{k_1}\delta^{\ell_2}_{k_2}\delta^{\ell_3}_{k_3}, \tag{5.1}$$

where S is the amplitude of the process shown in Fig. 2 and the star denotes the complex conjugation. If the second multiplier in the left-hand side of (5.1) is treated not as the complex-conjugated amplitude but the result of analytical continuation around the branching plane $\alpha=0$, the relation (5.1) becomes valid at any complex θ.

The requirement (5.1) for the two-colour string model leads, using (4.9), to the single equation for the "unitarizing factor" (4.10)

$$Z(\alpha,\beta,\gamma,\delta)Z^{(\alpha)}(\alpha,\beta,\gamma,\delta)=\frac{\cos\frac{\alpha+\beta}{2}\cos\frac{\alpha+\gamma}{2}\cos\frac{\alpha+\delta}{2}}{\cos\frac{\alpha}{2}\cos\frac{\beta}{2}\cos\frac{\gamma}{2}\cos\frac{\delta}{2}}, \tag{5.2}$$

where the suffix (α) denotes the continuation around the branching plane $\alpha=0$. This equation together with the requirement of symmetry under arbitrary

4 Actually, this is true unless the velocities of two-string intersection points exceed the speed of light

transmutation of the variables, determines the factor (4.10). The investigation of this equation is the subject of our further work.

It can be shown that, due to the factorization of the multistring S-matrix into three-string ones, the three-string unitarity condition (5.1) guarantees the unitarity of the total S-matrix of straight-strings.

6. Discussion

In a recent paper by Belavin [10] the remarkable symmetry of the triangle Eqs. (1.1) was discovered. This symmetry reveals the reasons for the compatibility of the overdefined system of functional Eqs. (1.1) and throws some light upon the nature of the general solution of these equations. It would be extremely interesting to find something like this symmetry in the tetrahedron equations.

As explained in the Introduction, the factorized S-matrix of straight-strings in the euclidean domain can be interpreted as the three-dimensional lattice statistical model which possesses Z-invariance and apparently is exactly solvable. Unfortunately, for the solution found in this paper some of the vertex weights turn out to be negative; therefore the existence of the thermodynamic limit of the corresponding lattice system becomes problematic. We suppose that there are solutions of tetrahedron equations which are free from this trouble. On the other hand, if the thermodynamic limit exists, there is the hypothesis that the partition function of a Z-invariant statistical system on the infinite lattice is simply connected to the "unitarizing factor" (4.10). (The two-dimensional analog of this hypothesis is discussed in [4].)

Appendix

Consider the Eq. (3.1) under the following condition

$$\theta_2+\theta_5+\theta_6=\pi. \tag{A.1}$$

Since $\mathcal{a}^{st}=0$, $S^{st}=\sigma^{st}=1$, the Eq. (3.1) acquires the form

$$\begin{aligned}&\mathcal{a}(\theta_1,\theta_2,\theta_3)\mathcal{a}(\theta_4,\pi-\theta_2-\theta_5,\theta_1)\mathcal{a}(\theta_4,\theta_3,\theta_5)\\&\quad=U(\theta_4,\theta_3,\theta_5)U(\theta_1,\theta_4,\pi-\theta_2-\theta_5)U(\theta_1,\theta_3,\theta_2).\end{aligned} \tag{A.2}$$

After the substitution of the explicit expressions (4.9) into (A.2) it can be rewritten

$$\begin{aligned}&\cos\frac{\theta_1+\theta_2+\theta_3}{2}\sin\frac{\theta_1+\theta_4-\theta_2-\theta_5}{2}\cos\frac{\theta_3+\theta_4+\theta_5}{2}\\&\quad=\cos\frac{\theta_4+\theta_3-\theta_5}{2}\sin\frac{\theta_1+\theta_2+\theta_4+\theta_5}{2}\cos\frac{\theta_1+\theta_3-\theta_2}{2}.\end{aligned} \tag{A.3}$$

The validity of this equality, assuming (3.2), remains to be proved.

The degeneration of the diagram in Fig. 7, corresponding to the case (A.1), is shown in Fig. 8. The length of the circular segment can be expressed independently in terms of the interior angles of two triangles in Fig. 8: either (123) or (124).

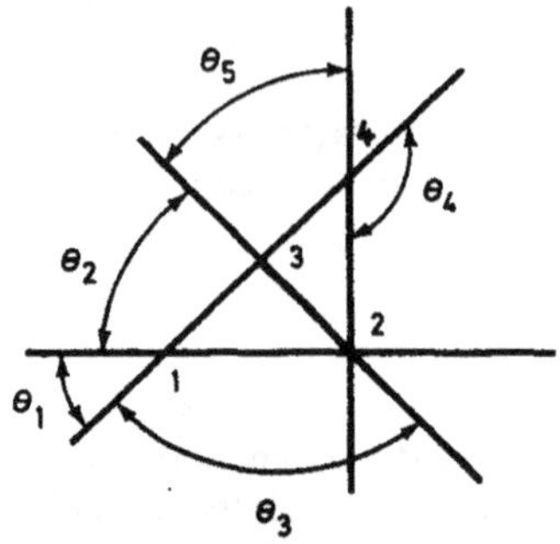

Fig. 8. Degeneration of diagram shown in Fig. 7, corresponding to the relation (A.1)

Comparing the results one obtains the relation

$$\left[\cos\frac{\theta_1+\theta_2+\theta_3}{2}\cos\frac{\theta_1+\theta_2-\theta_3}{2}\sin\frac{\theta_2+\theta_5-\theta_1-\theta_4}{2}\sin\frac{\theta_2+\theta_5+\theta_4-\theta_1}{2}\right]^{1/2}$$
$$=\left[\cos\frac{\theta_1+\theta_3-\theta_2}{2}\cos\frac{\theta_2+\theta_3-\theta_1}{2}\sin\frac{\theta_1+\theta_2+\theta_5-\theta_4}{2}\sin\frac{\theta_1+\theta_2+\theta_5+\theta_4}{2}\right]^{1/2}. \tag{A.4}$$

Doing the same with the segments ℓ_{34} and ℓ_{42} one gets two more relations

$$\left[\cos\frac{\theta_1+\theta_2+\theta_3}{2}\cos\frac{\theta_2+\theta_3-\theta_1}{2}\cos\frac{\theta_5+\theta_3-\theta_1}{2}\cos\frac{\theta_3+\theta_4+\theta_5}{2}\right]^{1/2}$$
$$=\left[\cos\frac{\theta_1+\theta_2-\theta_3}{2}\cos\frac{\theta_1+\theta_3-\theta_2}{2}\cos\frac{\theta_4+\theta_3-\theta_5}{2}\cos\frac{\theta_5+\theta_4-\theta_3}{2}\right]^{1/2}; \tag{A.5}$$

$$\left[\cos\frac{\theta_5+\theta_4-\theta_3}{2}\cos\frac{\theta_5+\theta_4+\theta_3}{2}\sin\frac{\theta_1+\theta_2+\theta_5-\theta_4}{2}\sin\frac{\theta_1+\theta_4-\theta_2-\theta_5}{2}\right]^{1/2}$$
$$=\left[\cos\frac{\theta_4+\theta_3-\theta_5}{2}\cos\frac{\theta_3+\theta_5-\theta_4}{2}\sin\frac{\theta_1+\theta_2+\theta_4+\theta_5}{2}\sin\frac{\theta_1-\theta_2-\theta_4-\theta_5}{2}\right]^{1/2}, \tag{A.6}$$

which are certainly equivalent to (A.4). Taking the products of the right- and left-hand sides of (A.4), (A.5), (A.6), one obtains exactly the equality (A.3).

Acknowledgements. I thank A. Belavin, V. Fateev, and Al. Zamolodchikov for many useful discussions. I am obliged to L. Shur, who has performed the computer verifications of the solution and to L. Agibalova for her invaluable help in preparing the English version of this article.

References

1. Yang, S.N.: Phys. Rev. **168**, 1920–1923 (1968)
2. Baxter, R.J.: Ann. Phys. **70**, 193–228 (1972)
3. Baxter, R.J.: Philos. Trans. Soc. (London) **189**, 315–346 (1978)
4. Zamolodchikov, A.B.: Sov. Sci. Rev. **2** (to be published)
5. Faddeev, L.D., Sklyanin, V.K., Takhtadjyan, L.A.: Teor. Mat. Fiz. **40**, 194–212 (1979)
6. Faddeev, L.D.: Preprint LOMI P-2-79, Leningrad, 1979
7. Zamolodchikov, A.B., Zamolodchikov, Al.B.: Ann. Phys. **120**, 253–291 (1979)
8. Zamolodchikov, A.B.: Landau Institute Preprint 15, Chernogolovka 1980, Zh. Eksp. Teor. Fiz. (in press)
9. Cherednik, I.: Dokl. Akad. Nauk USSR **249**, 1095–1098 (1979)
10. Belavin, A.A.: Pisma Zh. Eksp. Teòr. Fiz. (in press)

Communicated by Ya. G. Sinai

Received July 10, 1980

www.ingramcontent.com/pod-product-compliance
Lightning Source LLC
Chambersburg PA
CBHW060309220326
41598CB00027B/4283